STEEL DATA HANDBOOK

STEEL DATA HANDBOOK

K.V. Pramod
Professor
Department of Civil Engineering
SDM College of Engineering and Technology
Dhavalagiri, Dharwad, Karnataka

TECHSAR™
NEW DELHI

Published and Printed by
TechSar Pvt. Ltd.
4435-36/7, Ansari Road, Daryaganj,
New Delhi - 110 002 (INDIA)
E-mail: tech_info@techsarworld.com
Website: www.techsarworld.com

ISBN: 978-93-84588-34-2

Reprint 2025

Preface

he book is specially planned for reducing the time required for design of Steel Structural members. It gives the roperties of standard steel sections used in regular designs and also gives (i) Tensile Load capacity of Equal nd Unequal Angle Sections (ii) Compressive Strength and load capacity of single angle sections (both Equal and Jnequal Angles), all I-Sections, Channel Sections, T-Sections and (iii) Moment of Resistance of all I-sections. The etails of load capacities and Moments of Resistance are provided in the form of tables which are very easy to se. The tensile load capacities of angle sections are given for both yield stress and ultimate stress, for full cross ectional area and can be used for the design of welded members. The compressive load capacities are given for range of effective lengths starting from 1.0 m to an effective length corresponding to a slenderness ratio of 250, aximum value as per the code IS:800-2007. The Moments of Resistance for all plastic and compact I-sections are iven for both cases, viz., low shear and high shear, considering the compression flange to be laterally supported. he Moments of Resistance of semi-compact I-sections are given in a separate table, considering the compression ange to be laterally supported . The book also helps in making the designs economic as all sections having the esign load/moment capacities can be listed and the one with minimum weight per unit length can be selected s the design member. The book aims at increasing the efficiency of the steel structural designer – saving time nd simultaneously achieving economy. The students also can pick up this art while learning the subject.

K.V. Pramod

Symbols

a	Sectional Area
A	The longer leg of an unequal angle or one of the legs in case of an equal angle
b	Width of flange
B	The shorter leg of an unequal angle or one of the legs in case of an equal angle
C_{zz}	The lesser of the two extreme fibre distances from centroidal *zz* axis
C_{yy}	The lesser of the two extreme fibre distances from centroidal *yy* axis
e_{zz}	Distance of extreme fibre from the centroidal *zz* axis
e_{yy}	Distance of extreme fibre from the centroidal *yy* axis
g	Bolt gauge distance in the flange
g_1	Bolt gauge distance in web
h	Overall depth of section
h_1	Depth of the straight portion of the web
I_{zz}	Moment of Inertia about the centroidal *zz* axis
I_{yy}	Moment of Inertia about the centroidal *yy* axis
r_1	Radius at root of the flange
I_{zy}	Product of Inertia about the centroidal *zz* and *yy* axis
I_{uu}	Moment of Inertia about the centroidal *uu* axis
I_{vv}	Moment of Inertia about the centroidal *vv* axis
r_1	Radius at root of the flange
r_2	Radius at toe of the flange
r_{zz}	Radius of gyration about centroidal *zz* axis
r_{yy}	Radius of gyration about centroidal *yy* axis
r_{uu}	Radius of gyration about centroidal *uu* axis
r_{vv}	Radius of gyration about centroidal *vv* axis
t	Thickness of angles
t_f	Thickness of flange at the centre of the outstand
t_w	Thickness of web
w	Weight in *N* per meter length
Z_{zz}	Elastic modulus of section about centroidal *zz* axis
Z_{yy}	Elastic modulus of section about centroidal *yy* axis
Z_{pzz}	Plastic modulus of section about horizontal plastic neutral axis
Z_{pyy}	Plastic modulus of section about vertical plastic neutral axis
y_{pzz}	Distance of nearest extreme fibre from plastic neutral axis *zz*
y_{pyy}	Distance of nearest extreme fibre from plastic neutral axis *yy*
y-*y* axis	A line parallel to the axis of the web in case of beams, channels and Tee sections or parallel to the longer leg in case of unequal angles and passing through the section or equal area axis
z-*z* axis	A line passing through the centre of gravity of the section and at right angles to the *yy* axis
u-*u* axis	Axis through the centre of gravity representing major principal axis of angle section
v-*v* axis	Axis through the centre of gravity representing minor principal axis of angle section

D.C.S.	f_{cd} Design Compressive Strength
F.A.L.	Factored Axial Load
M_R	Moment of Resistance with low shear
M_{dV}	Moment of Resistance with high shear
V	Actual shear
V_d	Design shear strength
b	$((2V/V_d) - 1)^2$

Contents

'reface v
'ymbols vii

1. Rolled Steel I-Sections 2
2. Rolled Steel Channels 6
3. Rolled Steel Equal Angles 8
4. Rolled Steel Unequal Angle Sections 12
5. Rolled Steel Tee-Sections 15
6. Rolled Steel I-Sections with Additional Flange Plates to be used as Girders 17
7. Rolled Steel I-Sections with Channel and Plates on Flanges to be used as Girders 23

Tensile Load Capacity of Single Equal Angle Sections

8. Tensile Load Capacity of Equal Angle Sections for Yield Stress and Ultimate Stress With Welded end Connections for Gross Area 28

Tensile Load Capacity of Single Unequal Angle Sections

9. Tensile Load Capacity of Unequal Angle Sections for Yield Stress and Ultimate Stress with Welded end Connections for Gross Area 32

Factored Axial Compression Load Capacity of Single Equal Angle Section

10. Axial Compression Load Capacity of Single Equal Angle Section 36

Factored Axial Compression Load Capacity of Single Unequal Angle Section

11. Axial Compression Load Capacity of Single Unequal Angle Section 80

Factored Axial Compression Load Capacity of Single I-Sections

12. Axial Compression Load Capacity of I-Section 116

Factored Axial Compression Load Capacity of Single Channel-Sections

13. Axial Compression Load Capacity of Channel-Section 184

Factored Axial Compression Load Capacity of Single Tee-Sections

14. Axial Compression Load Capacity of T-Section 206

Factored Moment of Resistance of I-Sections

15. Moments of Resistance of Plastic and Compact I-Sections For Low Shear and High Shear 223
16. Moments of Resistance of Semi-compact I-Sections 224

Annexures

A-1 Shear Strength of Grade 4.6 Bolts 226
A-2 Tensile Strength of Grade 4.6 Bolts 226
A-3 Bearing Strength of Grade 4.6 Bolts for Typical end Distances and Pitch Distances 227
A-4 Fillet Weld Strength for Shop Welds and Field Welds 228

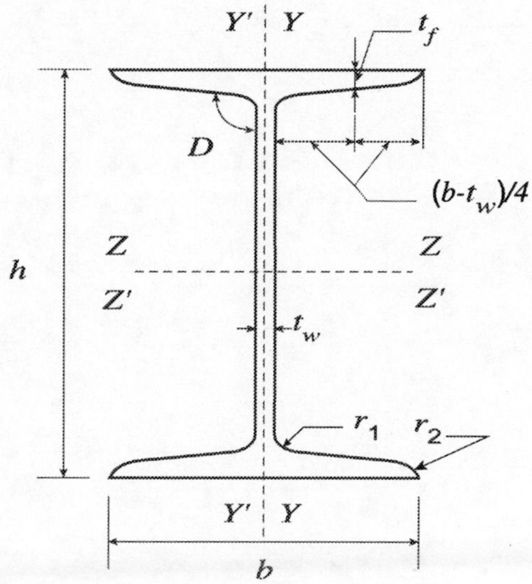

Figure 1 I-Sections

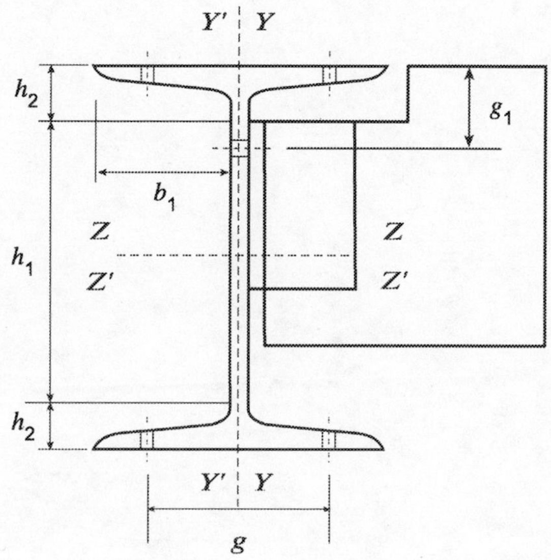

Figure 2 I-Sections

Table I Rolled Steel Beams

Designa-tion	Weight per metre	Sectional Area	Depth of section	Width of Flange	Thickness of Flange	Thickness of Web	Moments of inertia		Radii of Gyration		Radius at Root			Plastic Moduli of section		Section Classification						Designa-tion
	w	a	h	b	t_f	t_w	I_{zz}	I_{yy}	r_{zz}	r_{yy}	r_1	h_1	h_2	Zp_z	Zp_y	Flange	Web in		Entire section about		gf (Min)	
	in N	mm²	mm	mm	mm	mm	cm⁴	cm⁴	mm	mm	mm	mm	mm	mm³	mm³		Bending comp.	Axial comp.	z-z axis	y-y axis	mm	
ISJB 150	69.7	900.77	150	50	4.6	3.0	322.1	9.2	59.8	10.1	5.0	130.4	9.80	49573.97	5961.31	Plastic	Plastic	slender	*a*	*b*	45	ISJB 150
ISJB 175	79.5	1027.65	175	50	4.8	3.2	479.3	9.7	68.3	9.7	5.0	155	10.0	64223.19	6321.40	Plastic	Plastic	slender	*a*	*b*	45	ISJB 175
ISJB 200	97.1	1264.37	200	60	5.0	3.4	780.7	17.3	78.6	11.7	5.0	179.5	10.25	90897.85	9354.01	Plastic	Plastic	slender	*a*	*b*	45	ISJB 200
ISJB 225	125.6	1627.80	225	80	5.0	3.7	1308.5	40.5	89.7	15.8	6.5	201.1	11.95	134159.32	16291.29	Plastic	Plastic	slender	*a*	*b*	45	ISJB 225
ISLB 75	59.8	771.38	75	50	5.0	3.7	22.7	10.0	30.7	11.4	6.5	51.7	11.65	22352.00	6397.02	Plastic	Plastic	semi compact	*a*	*b*	–	ISLB 75
ISLB 100	78.5	1021.10	100	50	6.4	4.0	168	12.7	40.6	11.2	7.0	73	13.50	38986.81	8206.15	Plastic	Plastic	semi compact	*a*	*b*	50	ISLB 100
ISLB 125	116.7	1512.21	125	75	6.5	4.4	406.8	43.4	51.9	16.9	8.0	95.4	14.80	73931.87	18376.85	Plastic	Plastic	semi compact	*a*	*b*	50	ISLB 125
ISLB 150	139.3	1808.32	150	80	6.8	4.8	688.2	55.2	61.7	17.5	9.5	116.9	16.55	104507.89	22123.87	Plastic	Plastic	semi compact	*a*	*b*	50	ISLB 150
ISLB 175	163.8	2129.72	175	90	6.9	5.1	1096.2	79.6	71.7	19.3	9.5	141.6	16.70	142608.36	28342.30	Plastic	Plastic	semi compact	*a*	*b*	50	ISLB 175
ISLB 200	194.2	2526.76	200	100	7.3	5.4	1696.6	115.4	81.9	21.3	9.5	165.7	17.15	192946.92	36919.26	Plastic	Plastic	semi compact	*a*	*b*	50	ISLB 200
ISLB 225	230.5	2991.64	225	100	8.6	5.8	2501.9	112.7	91.5	19.4	12.0	180.3	22.35	254701.18	39221.19	Plastic	Plastic	semi compact	*a*	*b*	55	ISLB 225
ISLB 250	273.7	3552.89	250	125	8.2	6.1	3717.8	193.4	102.3	23.3	13.0	202.6	23.70	338671.31	55363.46	Plastic	Plastic	semi compact	*a*	*b*	60	ISLB 250
ISLB 275	323.7	4201.74	275	140	8.8	6.4	5375.3	287.0	113.1	26.1	14.0	223.7	25.65	443585.76	73535.39	Plastic	Plastic	semi compact	*a*	*b*	60	ISLB 275
ISLB 300	369.8	4807.79	350	150	9.4	6.7	7332.9	376.2	123.5	28	15.0	245.1	27.45	554301.90	89970.27	Plastic	Plastic	semi compact	*a*	*b*	60	ISLB 300
ISLB 325	422.8	5489.84	325	165	9.8	7.0	9847.6	510.8	134.1	30.5	16.0	266.5	29.25	687726.59	111868.67	Plastic	Plastic	semi compact	*a*	*b*	65	ISLB 325
ISLB 350	485.6	6301.32	350	165	11.4	7.4	13158.3	631.9	144.5	31.7	16.0	288.3	30.85	851096.20	134593.34	Plastic	Plastic	semi compact	*a*	*b*	65	ISLB 350
ISLB 400	558.2	7243.04	400	165	12.5	8.0	19305.3	716.4	163.3	31.5	16.0	336.2	31.90	1099441.34	151395.75	Plastic	Plastic	semi compact	*a*	*b*	65	ISLB 400
ISLB 450	640.6	8313.56	450	170	13.4	8.6	22536.1	853.0	182	32	16.0	384	33.0	1401285.36	174684.12	Plastic	Plastic	semi compact	*a*	*b*	70	ISLB 450
ISLB 500	735.8	9549.82	500	180	14.1	9.2	38579	1063.9	201	33.4	17.0	430.2	34.90	1772656.90	206685.84	Plastic	Plastic	semi compact	*a*	*b*	70	ISLB 500
ISLB 550	846.6	10997.40	550	190	15.0	9.9	53161.6	1335.1	219.9	34.8	18.0	476.1	36.95	2228166.68	246275.13	Plastic	Plastic	semi compact	*a*	*b*	70	ISLB 550
ISLB 600	976.1	12668.94	600	210	15.5	10.5	72867.6	1821.9	239.8	37.9	20.0	520.2	36.90	2798528.93	306808.35	Plastic	Plastic	semi compact	*a*	*b*	75	ISLB 600

Contd...

Designa-tion	Weight per metre	Sectional Area	Depth of section	Width of Flange	Thickness of Flange	Thickness of Web	Moments of inertia		Radii of Gyration		Radius at Root			Plastic Modulii of section		Section Classification						Designa-tion
	w	a	h	b	t_f	t_w	I_{zz}	I_{yy}	r_{zz}	r_{yy}	r_1	h_1	h_2	Zp_z	Zp_y	Flange	Web in		Entire section about		$g1$ (Min)	
	in N	mm^2	mm	mm	mm	mm	cm^4	cm^4	mm	mm	mm	mm	mm	mm^3	mm^3		Bending comp.	Axial comp.	z-z axis	y-y axis	mm	
ISMB 100	112.8	1459.75	100	75	7.2	4.0	257.5	40.8	42	16.7	9.0	65	17.50	58649.43	18221.39	Plastic	Plastic	semi compact	*a*	*b*	55	ISMB 100
ISMB 125	127.5	1660.47	125	75	7.6	4.4	449	43.7	52	16.2	9.0	89.2	17.90	81845.00	19580.61	Plastic	Plastic	semi compact	*a*	*b*	55	ISMB 125
ISMB 150	146.2	1900.39	150	80	7.6	4.8	726.4	52.6	61.8	16.6	9.0	113.9	18.05	110476.42	22319.85	Plastic	Plastic	semi compact	*a*	*b*	55	ISMB 150
ISMB 175	189.3	2462.01	175	90	8.5	5.5	1272	85.0	71.9	18.6	10.0	134.5	20.25	166076.07	32094.02	Plastic	Plastic	semi compact	*a*	*b*	55	ISMB 175
ISMB 200	240.2	3232.67	200	100	10.8	5.7	2235.4	150.0	83.2	21.5	11.0	152.7	23.65	253855.61	49988.90	Plastic	Plastic	semi compact	*a*	*b*	60	ISMB 200
ISMB 225	306.1	3971.50	225	100	11.8	6.5	3441.8	218.3	93.1	23.4	12.0	173.3	25.85	348270.85	66313.60	Plastic	Plastic	semi compact	*a*	*b*	60	ISMB 225
ISMB 250	365.9	4755.43	250	125	12.5	6.9	5131.6	334.5	10.39	2.65	13.0	194.1	27.95	465712.95	89701.12	Plastic	Plastic	semi compact	*a*	*b*	65	ISMB 250
ISMB 300	433.6	5626.38	300	140	12.4	7.5	8603.6	453.9	12.37	2.84	14.0	241.5	29.25	651730.75	110457.98	Plastic	Plastic	semi compact	*a*	*b*	65	ISMB 300
ISMB 350	514.0	6671.34	350	140	14.2	8.1	13630.3	537.7	14.29	2.84	14.0	288	31.00	889576.00	129723.39	Plastic	Plastic	semi compact	*a*	*b*	65	ISMB 350
ISMB 400	604.3	7845.58	400	140	16.0	8.9	20458.4	622.1	16.15	2.82	14.0	334.4	32.80	1176163.26	149666.89	Plastic	Plastic	semi compact	*a*	*b*	70	ISMB 400
ISMB 450	710.2	9226.63	450	150	17.4	9.4	30390.8	834.0	18.15	3.01	15.0	379.2	35.40	1553347.46	187128.80	Plastic	Plastic	semi compact	*a*	*b*	70	ISMB 450
ISMB 500	852.5	11074.38	500	180	17.2	10.2	45218.3	1369.8	20.21	3.52	17.0	424.1	37.95	2074650.52	259611.14	Plastic	Plastic	semi compact	*a*	*b*	75	ISMB 500
ISMB 550	1017	13211.08	550	190	19.3	11.2	64893.6	1833.8	22.16	3.73	18.0	467.5	41.25	2711958.11	328051.25	Plastic	Plastic	semi compact	*a*	*b*	75	ISMB 550
ISMB 600	1210	15621.24	600	210	20.3	12.0	91813.0	2651.0	24.24	4.11	20.0	509.7	45.15	3510597.39	429318.06	Plastic	Plastic	semi compact	*a*	*b*	80	ISMB 600
ISWB 150	167	2166.53	150	100	7.0	5.4	839.1	94.8	6.22	2.09	8.0	116.6	16.70	126857.02	31938.83	Plastic	Plastic	semi compact	*a*	*b*	55	ISWB 150
ISWB 175	217	2811.29	175	125	7.4	5.8	1509.4	188.6	7.33	2.59	8.0	139.5	17.75	194189.53	51272.20	Plastic	Plastic	semi compact	*a*	*b*	55	ISWB 175
ISWB 200	283	3670.87	200	140	9.0	6.1	2624.5	328.8	8.46	2.99	9.0	158.8	20.60	293934.76	78702.39	Plastic	Plastic	semi compact	*a*	*b*	55	ISWB 200
ISWB 225	333	4323.95	225	150	9.9	6.4	3920.5	448.6	9.52	3.22	9.0	181.4	21.80	389930.93	99771.54	Plastic	Plastic	semi compact	*a*	*b*	55	ISWB 225
ISWB 250	401	5204.61	250	200	9.0	6.7	5943.1	875.5	106.9	40.6	10.0	203.8	23.10	527556.03	149659.92	Semi Compact	Plastic	semi compact	*a*	*b*	60	ISWB 250
ISWB 300	472	6132.75	300	200	10.0	7.4	9821.6	990.1	126.6	40.2	11.0	250.1	24.95	731207.54	171014.22	Compact	Plastic	semi compact	*a*	*b*	60	ISWB 300

Contd...

Designa-tion	Weight per metre	Sectional Area	Depth of section	Width of Flange	Thickness of Flange	Thickness of Web	Moments of inertia		Radii of Gyration		Radius at Root			Plastic Moduli of section		Section Classification						Designa-tion
	w	a	h	b	t_f	t_w	I_{zz}	I_{yy}	r_{zz}	r_{yy}	r_1	h_1	h_2	Zp_z	Zp_y	Flange	Web in		Entire section about		g1 (Min)	
	in N	mm^2	mm	mm	mm	mm	cm^4	cm^4	mm	mm	mm	mm	mm	mm^3	mm^3		Bending comp.	Axial comp.	z-z axis	y-y axis	mm	
ISWB 350	558	7249.90	350	200	11.4	8.0	16521.7	1175.9	146.3	40.3	12.0	295.5	27.25	895484.03	200466.66	Plastic	Plastic	semi compact	a	b	60	ISWB 350
ISWB 400	654	8501.25	400	200	13.0	8.6	23426.7	1388.0	166	40.4	13.0	340.5	29.75	1320176.95	234187.39	Plastic	Plastic	semi compact	a	b	65	ISWB 400
ISWB 450	779	10115.05	450	200	15.4	9.2	35057.6	1706.7	186.3	41.1	14.0	384	33.00	1760576.17	284175.38	Plastic	Plastic	semi compact	a	b	70	ISWB 450
ISWB 500	934	12121.91	500	250	14.7	9.9	52290.9	2987.8	207.7	49.6	15.0	431	34.50	2353350.98	406821.59	Plastic	Plastic	semi compact	a	b	70	ISWB 500
ISWB 550	1104	14333.94	550	250	17.6	10.5	74906.1	3740.6	228.6	51.1	16.0	473.4	38.30	3066252.60	500169.11	Plastic	Plastic	semi compact	a	b	75	ISWB 550
ISWB 600	1312	17037.99	600	250	21.3	11.2	106198.5	4702.5	249.7	52.5	17.0	514.2	42.90	3986643.14	619223.64	Plastic	Plastic	semi compact	a	b	80	ISWB 600
ISWB 600	1423	18514.04	600	250	23.6	11.8	115626.6	5298.3	250.1	53.5	18.0	507.9	46.05	4349740.84	695987.82	Plastic	Plastic	semi compact	a	b	80	ISWB 600
ISHB 150	266	3447.75	150	150	9.0	5.4	1455.6	431.7	65	35.4	8.0	112	19.00	215643.31	92741.44	Plastic	Plastic	semi compact	b	c	55	ISHB 150
ISHB 150	300	3897.75	150	150	9.0	8.4	1540.0	460.3	62.9	34.4	8.0	112	19.00	232518.31	98250.56	Plastic	Plastic	semi compact	b	c	55	ISHB 150
ISHB 150	339	4407.75	150	150	9.0	11.8	1635.6	494.9	60.9	33.5	8.0	112	19.00	251643.31	105310.23	Plastic	Plastic	semi compact	b	c	55	ISHB 150
ISHB 200	366	4754.43	200	200	9.0	6.1	3608.4	967.1	87.1	45.1	9.0	158.4	20.80	397232.11	159280.82	Semi Comp.	Plastic	semi compact	b	c	55	ISHB 200
ISHB 200	392	5094.43	200	200	9.0	7.8	3721.8	994.6	85.5	44.2	9.0	158.4	20.80	414232.11	163486.59	Semi Comp.	Plastic	semi compact	b	c	55	ISHB 200
ISHB 225	423	5493.81	225	225	9.1	6.5	5279.5	1353.8	98	49.6	10.0	180.5	22.20	515629.31	200491.52	Semi Comp.	Plastic	semi compact	b	c	55	ISHB 225
ISHB 225	459	5966.31	225	225	9.1	8.6	5478.8	1396.6	95.8	48.4	10.0	180.5	22.20	542207.43	206508.08	Semi Comp.	Plastic	semi compact	b	c	55	ISHB 225
ISHB 250	500	6495.75	250	250	9.7	6.9	7736.5	1961.3	109.1	54.9	10.0	203.5	23.20	678745.47	262155.26	Semi Comp.	Plastic	semi compact	b	c	60	ISHB 250
ISHB 250	537	6970.75	250	250	9.7	8.8	7983.9	2011.7	107	53.7	10.0	203.5	23.20	708432.97	268551.85	Semi Comp.	Plastic	semi compact	b	c	60	ISHB 250
ISHB 300	577	7484.95	300	250	10.6	7.6	12545.2	2193.6	129.5	54.1	11.0	249.8	25.10	921680.00	291583.52	Semi Comp.	Plastic	semi compact	b	c	60	ISHB 300
ISHB 300	618	8024.95	300	250	10.6	9.4	12950.2	2246.7	127	52.9	11.0	249.8	25.10	962180.00	298582.93	Semi Comp.	Plastic	semi compact	b	c	60	ISHB 300
ISHB 350	661	8591.07	350	250	11.6	8.3	19159.7	2451.4	149.3	53.4	12.0	296	27.00	1213551.49	324438.35	Semi Comp.	Plastic	semi compact	a	b	60	ISHB 350
ISHB 350	710	9221.07	350	250	11.6	10·j	19802.8	2510.5	146.5	52.2	12.0	296	27.00	1268876.49	332453.82	Semi Comp.	Plastic	semi compact	a	b	60	ISHB 350

Contd...

tion	per metre	Area	of section	of Flange	of Flange	of Web	of inertia		Radii of Gyration		at Root			Plastic modulus of section		Section Classification						Designa-tion
	w	a	h	b	t_f	t_w	I_{zz}	I_{yy}	r_{zz}	r_{yy}	r_1	h_1	h_2	Zp_z	Zp_y	Flange	Web in		Entire section about		g1 (Min)	
	in N	mm^2	mm	mm	mm	mm	cm^4	cm^4	mm	mm	mm	mm	mm	mm^3	mm^3		Bending comp.	Axial comp.	z-z axis	y-y axis	mm	
ISHB 400	759	9865.89	400	250	12.7	9.1	28083.5	2728.3	168.7	52.6	14.0	340.1	29.90	1566357.36	360549.61	Compact	Plastic	semi compact	a	b	65	ISHB 400
ISHB 400	806	10465.89	400	250	12.7	10.6	28823.5	2783.0	166.1	51.6	14.0	340.1	29.90	1626357.36	368174.02	Compact	Plastic	semi compact	a	b	65	ISHB 400
ISHB 450	859	11114.35	450	250	13.7	9.8	39210.8	2985.2	187.8	51.8	15.0	386.2	31.90	1955034.44	394145.02	Plastic	Plastic	semi compact	a	b	65	ISHB 450
ISHB 450	907	11789.35	450	250	13.7	11.3	40349.9	3045.0	185	50.8	15.0	386.2	31.90	2030971.94	402733.91	Plastic	Plastic	semi compact	a1	b	65	ISHB 450

$$Z_{zz} = \frac{I_{zz}}{h/2},\ Z_{yy} = \frac{I_{yy}}{b/2}$$

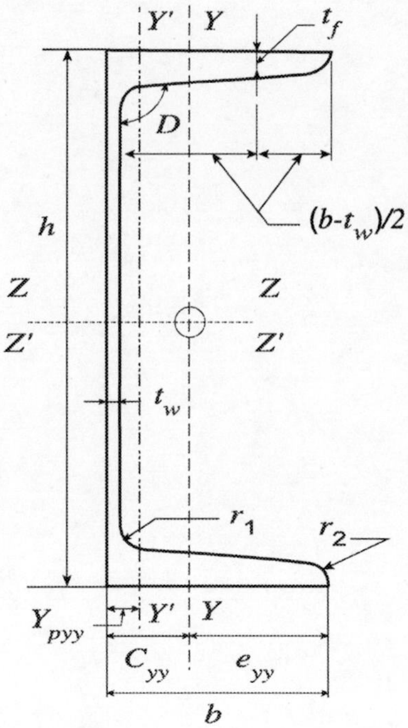

Figure 3 Channels

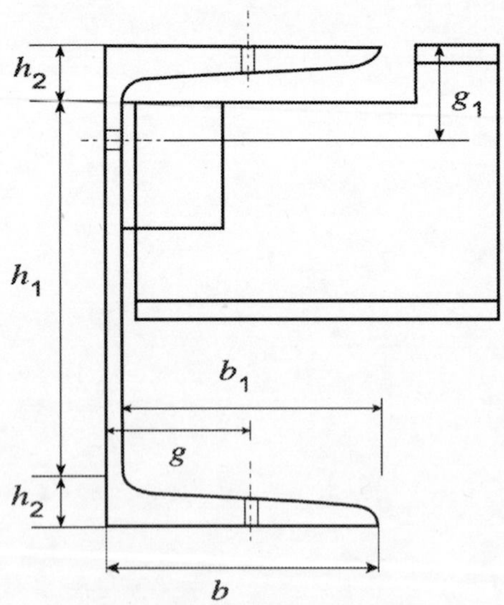

Figure 4 Channels

Designation	Weight per metre	Sectional Area	Depth of section	Width of Flange	Thickness of Flange	Thickness of Web	Centre of Gravity	Moments of inertia		Distance of Plastic N-A	Plastic Moduli of section		Radii of Gyration		Radius at Root	Connection details				Buckling Class				Designation
	w	a	h	b	t_f	t_w	C_{yy}	I_{zz}	I_{yy}	yp_{yy}	Zp_{zz}	Zp_{yy}	r_{zz}	r_{yy}	r_1	h_1	h_2	g	$g1$ (Min)			Entire Section		
	N	mm²	mm	mm	mm	mm	mm	cm⁴	cm⁴	mm	mm³	mm³	mm	mm	mm	mm	mm	mm	mm	Flange	Web	About z – z axis	About y – y axis	
ISJC 100	57	741.32	100	45	5.1	3.0	14.0	123.8	14.9	6.67	28375.84	8807.79	40.9	14.2	6.0	77	11.5	25	50	Plastic	Plastic	c	c	ISJC 100
ISJC 125	77	1007.41	125	50	6.6	3.0	16.4	270.0	25.7	11.22	49082.07	13640.64	51.8	16.0	6.0	98.9	13.1	28	50	Plastic	Plastic	c	c	ISJC 125
ISJC 150	97	1265.47	150	55	6.9	3.6	16.6	471.1	37.9	7.30	72047.18	18104.60	61.0	17.3	7.0	121.2	14.4	30	60	Plastic	Plastic	c	c	ISJC 150
ISJC 175	110	1424.47	175	60	6.9	3.6	17.5	719.9	50.5	6.79	94223.46	21865.41	71.1	18.8	7.0	146.1	14.5	35	50	Plastic	Plastic	c	c	ISJC 175
ISJC 200	136	1776.67	200	70	7.1	4.1	19.7	1161.2	84.2	6.45	133107.66	30864.13	80.8	21.8	8.0	168.5	15.8	40	50	Compact	Plastic	c	c	ISJC 200
ISLC 75	56	726.02	75	40	6.0	3.7	13.5	66.1	11.5	7.01	20609.19	7802.10	30.2	12.6	6.0	50.4	12.3	21	–	Plastic	Plastic	c	c	ISLC 75
ISLC 100	77	1001.72	100	50	6.4	4.0	16.2	164.7	24.8	10.34	38085.13	13288.61	40.6	15.7	6.0	74.3	12.8	28	50	Plastic	Plastic	c	c	ISLC 100
ISLC 125	105	1367.18	125	65	6.6	4.4	20.4	356.8	57.2	12.39	65447.21	23410.59	51.1	20.5	7.0	96.6	14.2	35	50	Compact	Plastic	c	c	ISLC 125
ISLC 150	141	1836.01	150	75	7.8	4.8	23.8	697.2	103.2	15.12	106172.65	36552.18	61.6	23.7	8.0	117.0	16.5	40	50	Compact	Plastic	c	c	ISLC 150
ISLC 175	173	2239.98	175	75	9.5	5.1	24.0	1148.4	126.5	15.22	150361.55	44744.64	71.6	23.8	8.0	138.6	18.2	40	55	Plastic	Plastic	c	c	ISLC 175
ISLC 200	202	2622.19	200	75	10.8	5.5	23.5	1725.5	146.9	11.79	198774.13	51864.13	81.1	23.7	8.5	160.0	20.0	40	55	Plastic	Plastic	c	c	ISLC 200
ISLC 225	235	3053.06	225	90	10.2	5.8	24.6	2547.9	209.5	10.91	260128.43	63746.28	91.4	26.2	11.0	175.9	24.5	50	60	Plastic	Plastic	c	c	ISLC 225
ISLC 250	275	3564.84	250	100	10.7	6.1	27.0	3687.9	2984.0	11.87	338113.80	82129.70	101.7	28.9	11.0	198.6	25.5	60	60	Plastic	Plastic	c	c	ISLC 250
ISLC 300	325	4210.71	300	100	11.6	6.7	25.5	6047.9	346.0	8.52	466722.08	92494.64	119.8	28.7	12.0	245.4	27.3	60	60	Plastic	Plastic	c	c	ISLC 300
ISLC 350	381	4949.52	350	100	12.5	7.4	24.1	9312.6	394.6	7.07	623304.66	101861.17	137.2	28.2	13.0	291.9	29.1	60	65	Plastic	Slender	c	c	ISLC 350
ISLC 400	448	5825.21	400	100	14.0	8.0	23.6	13989.5	460.4	7.28	825020.00	116140.39	155.0	28.1	14.0	337.1	31.4	60	65	Plastic	Plastic	c	c	ISLC 400
ISMC 75	67	867.17	75	40	7.3	4.4	13.1	76.0	12.6	7.88	24164.88	8636.28	29.6	12.1	8.5	41.4	16.8	21	–	Plastic	Plastic	c	c	ISMC 75
ISMC 100	90	1169.83	100	50	7.5	4.7	15.3	187.7	25.9	8.29	43825.20	14208.25	40.0	14.9	9.0	64.0	18.0	28	50	Plastic	Plastic	c	c	ISMC 100
ISMC 125	125	1618.84	125	65	8.1	5.0	19.4	416.4	59.9	10.51	77152.85	25462.96	50.7	19.2	9.5	85.4	19.8	35	55	Plastic	Plastic	c	c	ISMC 125
ISMC 150	161	2087.90	150	75	9.0	5.4	22.2	779.4	102.3	11.87	119822.80	37826.78	61.1	22.1	10.0	106.7	21.7	40	55	Plastic	Plastic	c	c	ISMC 150
ISMC 175	187	2438.00	175	75	10.2	5.7	22.0	1223.3	121.0	11.17	161643.32	44151.63	70.8	22.3	10.5	128.4	23.3	40	55	Plastic	Plastic	c	c	ISMC 175
ISMC 200	217	2821.30	200	75	11.4	6.1	21.7	1819.3	140.4	10.32	211254.07	50724.65	80.8	22.3	11.0	150.2	24.9	40	60	Plastic	Plastic	c	c	ISMC 200
ISMC 225	254	3301.43	225	80	12.4	6.4	23.0	2694.6	187.2	10.66	277928.22	63179.08	90.3	23.8	12.0	170.9	27.1	45	60	Plastic	Plastic	c	c	ISMC 225
ISMC 250	293	3866.93	250	80	14.1	7.1	23.0	3816.8	219.1	10.03	356720.73	73345.89	99.4	26.1	12.0	192.5	28.7	45	65	Plastic	Plastic	c	c	ISMC 250
ISMC 300	351	4563.71	300	90	13.6	7.6	23.6	6362.6	310.8	7.63	496767.49	90581.08	118.1	28.3	13.0	240.7	29.6	50	65	Plastic	Plastic	c	c	ISMC 300
ISMC 350	413	5365.51	350	100	13.5	8.1	24.4	10008.0	430.6	7.67	672190.45	110528.21	136.6	28.3	14.0	286.1	30.9	60	65	Plastic	Plastic	c	c	ISMC 350
ISMC 400	489	6293.33	400	100	15.3	8.6	24.2	15082.8	504.8	7.87	891020.16	127542.61	154.8	28.3	15.0	332.8	33.6	60	70	Plastic	Plastic	c	c	ISMC 400

$$Z_{zz} = \frac{I_{zz}}{h/2},\ Z_{yy} = \frac{I_{yy}}{b/2}$$

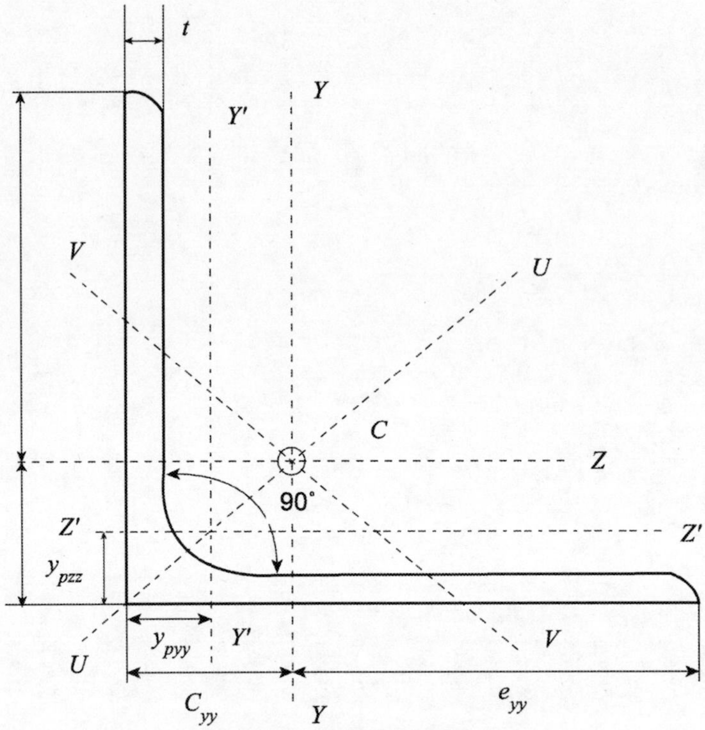

Figure 5 Equal Angles

Designation	Thickness	Sectional Area	Weight per metre	Centre of Gravity	Distance of Extreme Fibre	Moments of Inertia			Radii of Gyration			Moduli of section	Distance of Plastic N-A	Plastic Modulus	Radius at Root	Product of inertia	Section Classification		Designation
	t	A	N	$C_{zz} = C_{yy}$	$e_{zz} = e_{yy}$	$I_{zz} = I_{yy}$	I_{uu}	I_{vv}	$r_{zz} = r_{yy}$	r_{uu}	r_{vv}	$Z_{zz} = Z_{yy}$	$Yp_{zz} = YP_{yy}$	$Zp_{zz} = Zp_{yy}$	r_1	I_{zy}	Bending Comp.	Axial Comp.	
	mm	mm^2		mm	mm	cm^4	cm^4	cm^4	mm	mm	mm	cm^3	mm	mm^3	mm	cm^4			
ISA 2020	3.0	111.75	8.6	5.9	14.1	0.4	0.6	0.2	5.8	7.3	3.7	0.3	2.8	502.55	4.0	0.2			ISA 2020
	4.0	144.75	11.1	6.3	13.7	0.5	0.8	0.2	5.8	7.2	3.7	0.4	3.6	650.97	4.0	0.3			
ISA 2525	3.0	141.48	10.8	7.1	17.9	0.8	1.2	0.3	7.3	9.3	4.7	0.4	2.9	798.90	4.5	0.4			ISA 2525
	4.0	184.48	14.2	7.5	17.5	1	1.6	0.4	7.3	9.1	4.7	0.6	3.7	1043.03	4.5	0.6			
	5.0	225.48	17.3	7.9	17.1	1.2	1.8	0.5	7.2	9.1	4.7	0.7	4.5	1272.93	4.5	0.7			
ISA 3030	3.0	172.50	13.2	8.3	21.7	1.4	2.2	0.6	8.9	11.3	5.7	0.6	2.9	1180.86	5.0	0.8			ISA 3030
	4.0	225.50	17.3	8.7	21.3	1.8	2.8	0.7	8.9	11.2	5.7	0.8	3.8	1545.15	5.0	1.0			
	5.0	276.50	21.2	9.2	20.8	2.1	3.4	0.9	8.8	11.1	5.7	1.0	4.6	1890.48	5.0	1.2			
ISA 3535	3.0	202.50	15.5	9.5	25.5	2.3	3.6	0.9	10.5	13.3	6.7	0.9	2.9	1636.18	5.0	1.3			ISA 3535
	4.0	265.50	20.4	10.0	25.0	2.9	4.7	1.2	10.5	13.2	6.7	1.2	3.8	2145.50	5.0	1.7			
	5.0	326.50	25.1	10.4	24.6	3.5	5.6	1.5	10.4	13.1	6.7	1.4	4.7	2631.01	5.0	2.1			
	6.0	385.50	29.6	10.8	24.2	4.1	6.5	1.7	10.3	12.9	6.7	1.7	5.5	3094.99	5.0	2.4			
ISA 4040	3.0	233.63	17.9	10.8	29.2	3.4	5.5	1.4	12.1	15.4	7.7	1.2	3.0	2168.49	5.5	2.0			ISA 4040
	4.0	306.63	23.6	11.2	28.8	4.5	7.1	1.8	12.1	15.3	7.7	1.6	3.9	2847.91	5.5	2.6			
	5.0	377.63	29.0	11.6	28.4	5.4	8.6	2.2	12.0	15.1	7.7	1.9	4.7	3498.66	5.5	3.2			
	6.0	446.63	34.3	12.0	28.0	6.3	10.0	2.6	11.9	15.0	7.7	2.3	5.6	4123.13	5.5	3.7			
ISA 4545	3.0	263.63	20.3	12.0	33.0	5.0	8.0	2.0	13.8	17.4	8.7	1.5	3.0	2773.82	5.5	2.9			ISA 4545
	4.0	346.63	26.6	12.5	32.5	6.5	10.4	2.6	13.7	17.3	8.7	2.0	3.9	3648.25	5.5	3.8			
	5.0	427.63	32.9	12.9	32.1	7.9	12.6	3.2	13.6	17.2	8.7	2.5	4.8	4489.08	5.5	4.6			
	6.0	506.63	39.0	13.3	31.7	9.2	14.6	3.8	13.5	17.0	8.7	2.9	5.6	5298.79	5.5	5.4			
ISA 5050	3.0	294.86	22.7	13.2	36.8	6.9	11.1	2.8	15.3	19.4	9.7	1.9	3.0	3456.51	6.0	4.1			ISA 5050
	4.0	387.86	29.8	13.7	36.3	9.1	14.5	3.6	15.3	19.3	9.7	2.5	3.9	4551.02	6.0	5.3			
	5.0	478.86	37.3	14.1	35.9	11.0	17.6	4.5	15.2	19.2	09.7	3.10	4.82	5607.02	6.00	6.50	Compact	Semi comp.	
	6.0	567.86	44.1	14.5	35.5	12.9	20.6	5.3	15.1	19.0	09.6	3.60	5.70	6627.04		7.60	Plastic	Semi comp.	
ISA 5555	5.0	527.20	40.2	15.3	39.7	14.7	23.5	5.9	16.7	21.1	10.6	3.70	4.84	6773.52	6.50	8.60	Semi comp.	Semi comp.	ISA 5555
	6.0	626.20	48.1	15.7	39.3	17.3	27.5	7.0	16.6	21.0	10.6	4.40	5.74	8029.84		10.10	Plastic	Semi comp.	
	8.0	818.20	62.8	16.5	38.5	22.0	34.9	9.1	16.4	20.7	10.6	5.70	7.46	10430.02		12.80	Plastic	Semi comp.	
	10.0	1002.20	77.5	17.2	37.8	26.3	41.5	11.2	16.2	20.3	10.6	7.00	9.11	12695.32		15.10	Plastic	Semi comp.	
ISA 6060	5.0	575.38	44.1	16.5	43.5	19.2	30.6	· 7.7	18.2	23.1	11.6	4.40	4.86	8080.31	6.50	11.30	Semi comp.	Semi comp.	ISA 6060
	6.0	684.38	53	16.9	43.1	22.6	36.0	9.1	18.2	22.9	11.5	5.20	5.76	9597.31		13.30	Compact	Semi comp.	
	8.0	896.38	68.7	17.7	42.3	29.0	46.0	11.9	18.0	22.7	11.5	6.80	7.51	12504.41		16.90	Plastic	Semi comp.	
	10.0	1100.38	84.4	18.5	41.5	34.8	54.9	14.6	17.8	22.3	11.5	8.40	9.18	15257.92		20.10	Plastic	Semi comp.	
ISA 6565	5.0	625.38	48.1	17.7	47.3	24.7	39.4	9.9	199	25.1	12.6	5.20	4.87	9558.57	6.50	14.50	Semi comp.	Slender	ISA 6565
	6.0	744.38	56.9	18.1	46.9	29.1	46.5	11.7	19.8	25.0	12.6	6.20	5.78	11360.62		17.20	Semi comp.	Semi comp.	

Contd...

Designation	Thickness	Sectional Area	Weight per metre	Centre of Gravity	Distance of Extreme Fibre	Moments of Inertia			Radii of Gyration			Moduli of section	Distance of Plastic N-A	Plastic Modulus	Radius at Root	Product of inertia	Section Classification		Designation
	t	A	N	$C_{zz} = C_{yy}$	$e_{zz} = e_{yy}$	$I_{zz} = I_{yy}$	I_{uu}	I_{vv}	$r_{zz} = r_{yy}$	r_{uu}	r_{vv}	$Z_{zz} = Z_{yy}$	$Yp_{zz} = YP_{yy}$	$Zp_{zz} = Zp_{yy}$	r_1	I_{zy}	Bending Comp.	Axial Comp.	
	mm	mm²		mm	mm	cm⁴	cm⁴	cm⁴	mm	mm	mm	cm³	mm	mm³	mm	cm⁴			
	8.0	976.38	75.5	18.9	46.1	37.4	59.5	15.3	19.6	24.7	12.5	8.10	7.55	14823.16		22.00	Plastic	Semi comp	
	10.0	1200.38	92.2	19.7	45.3	45.0	71.3	18.8	19.4	24.4	12.5	9.90	9.25	18112.90		26.20	Plastic	Semi comp	
ISA 7070	5.0	676.82	52.0	18.9	51.1	31.1	49.8	12.5	21.5	27.1	13.6	6.10	4.89	11165.18	7.00	18.40	Semi comp	Slender	ISA 7070
	6.0	805.82	61.6	19.4	50.6	36.8	58.8	14.8	21.4	27	13.6	7.3	5.81	13277.37		21.7	Semi comp.	Semi comp.	
	8.0	1057.82	81.4	20.2	49.8	47.4	75.5	19.3	21.2	26.7	13.5	9.5	7.59	17345.54		27.9	Plastic	Semi comp.	
	10.0	1301.82	100.1	21.0	49.0	57.2	90.7	23.7	21	26.4	13.5	11.7	9.32	21221.66		33.3	Plastic	Semi comp.	
ISA 7575	5.0	726.82	55.9	20.2	54.8	38.7	61.9	15.5	23.1	29.2	14.6	7.10	4.90	12893.44	7.00	22.80	Semi comp.	Slender	ISA 7575
	6.0	865.82	66.7	20.6	54.4	45.7	73.1	18.4	23	29.1	14.6	8.4	5.82	15340.66		27.0	Semi comp.	Semi comp.	
	8.0	1137.82	87.3	21.4	53.6	59.0	94.1	24	22.8	28.8	14.5	11.0	7.62	20064.13		34.8	Plastic	Semi comp.	
	10.0	1401.82	107.9	22.2	52.8	71.4	113.3	29.4	22.6	28.4	14.5	13.5	9.37	24576.12		41.7	Plastic	Semi comp.	
ISA 8080	6.0	929.04	71.6	21.8	58.2	56.0	89.6	22.5	24.6	31.1	15.6	9.60	5.85	17562.32	8.00	33.00	Semi comp.	Slender	ISA 8080
	8.0	1221.04	94.2	22.7	57.3	72.5	115.6	29.4	24.4	30.8	15.5	12.6	7.67	22991.54		42.7	Compact	Semi comp.	
	10.0	1505.04	115.8	23.4	56.6	87.7	139.5	36	24.1	30.4	15.5	15.5	9.43	28189.96		51.4	Plastic	Semi comp	
	12.0	1781.04	137.3	24.2	55.8	101.9	161.4	42.4	23.9	30.1	15.4	18.3	11.14	33176.56		59.2	Plastic	Semi comp	
ISA 9090	6.0	1046.52	80.4	24.2	65.8	80.1	128.1	32.0	27.7	35.0	17.5	12.20	5.88	22261.00	3.50	47.20	Semi comp	Slender	ISA 9090
	8.0	1378.52	105.9	25.1	64.9	104.2	166.4	42.0	27.5	34.7	17.5	16.0	7.71	29253.55		61.5	Semi comp.	Semi comp.	
	10.0	1702.52	131.5	25.9	64.1	126.7	201.9	51.6	27.3	34.4	17.4	19.8	9.50	35976.62		74.5	Plastic	Semi comp.	
	12.0	2018.52	155.0	26.6	63.4	147.9	234.9	60.9	27.1	34.1	17.4	23.3	11.24	42449.88		86.5	Plastic	Semi comp.	
ISA 100100	6.0	1166.52	90.2	26.7	73.3	111.3	178.1	44.5	30.9	39.1	19.5	15.20	5.89	27716.04	8.50	65.70	Slender	Slender	ISA 100100
	8.0	1538.52	118.7	27.6	72.4	145.1	231.8	58.4	30.7	38.8	19.5	20.0	7.74	36468.80		85.8	Semi comp.	Semi comp.	
	10.0	1902.52	146.2	28.4	71.6	177.0	282.2	71.8	30.5	38.5	19.4	24.7	9.55	44912.66		104.4	Compact	Semi comp.	
	12.0	2258.52	173.6	29.2	70.8	207.0	329.3	84.7	30.3	38.2	19.4	29.2	11.32	53067.74		121.6	Plastic	Semi comp.	
ISA 110110	8.0	1702.01	131.50	30.0	80.0	195.0	311.7	78.2	33.8	42.8	21.4	24.4	7.80	44375.14	10.00	115.1	Semi comp.	Slender	ISA 110110
	10.0	2106.01	161.90	30.8	79.2	238.4	380.5	96.3	33.6	42.5	21.4	30.1	9.62	54742.14		140.6	Semi comp.	Semi comp.	
	12.0	2502.01	192.30	31.6	78.4	279.6	445.3	113.8	33.4	42.2	21.3	35.7	11.41	64781.66		164.5	Plastic	Semi comp.	
	15.0	3081.01	237.40	32.7	77.3	337.4	535.4	139.3	33.1	41.7	21.3	43.7	14.02	79269.58		197.0	Plastic	Semi comp	
ISA 130130	8.0	2022.01	156.0	35.0	95.0	328.3	525.1	131.4	40.3	51.0	25.5	34.50	7.83	62780.66	10.00	194.20	Slender	Slender	ISA 130130
	10.0	2506.01	193.30	35.8	94.2	402.7	643.4	162.1	40.1	50.7	25.4	42.7	9.68	77588.42		238.3	Semi comp.	Slender	
	12.0	2982.01	229.60	36.6	93.4	473.8	755.9	191.8	39.9	50.3	25.4	50.7	11.51	91989.91		279.9	Semi comp.	Semi comp.	
	15.0	3681.01	283.50	37.8	92.2	574.6	914.2	235	39.5	49.8	25.3	62.3	14.18	112874.63		337.8	Plastic	Semi comp.	
ISA 150150	10.0	2903.43	223.7	40.6	109.4	622.4	995.4	249.4	46.3	58.6	29.3	56.90	9.76	103615.53	12.00	368.20	Semi comp.	Slender	ISA 150150
	12.0	3459.43	266.80	41.4	108.6	735.4	1174.8	296	46.1	58.3	29.3	67.7	11.60	123187.47		435.0	Semi comp.	Semi comp.	
	15.0	4278.43	329.60	42.6	107.4	896.8	1429.7	363.8	45.8	57.8	29.2	83.5	14.32	151682.69		529.1	Compact	Semi comp.	
	18.0	5079.43	391.40	43.8	106.2	1048.9	1668.2	429.5	45.4	57.3	29.1	98.7	16.97	179202.78		616.0	Plastic	Semi comp.	
ISA 200200	12.0	4661.37	359.0	53.6	146.4	1788.9	2862.0	715.9	62.0	78.4	39.2	122.20	11.75	222510.80	15.00	1058.90	Slender	Slender	ISA 200200
	15.0	5780.37	445.40	54.9	145.1	2197.7	3511.8	883.7	61.7	77.9	39.1	151.4	14.53	275254.93		1301.2	Semi comp.	Slender	
	18.0	6881.37	529.70	56.1	143.9	2588.7	4130.8	1046.5	61.3	77.5	39	179.9	17.27	326584.31		1530.5	Semi comp.	Semi comp	

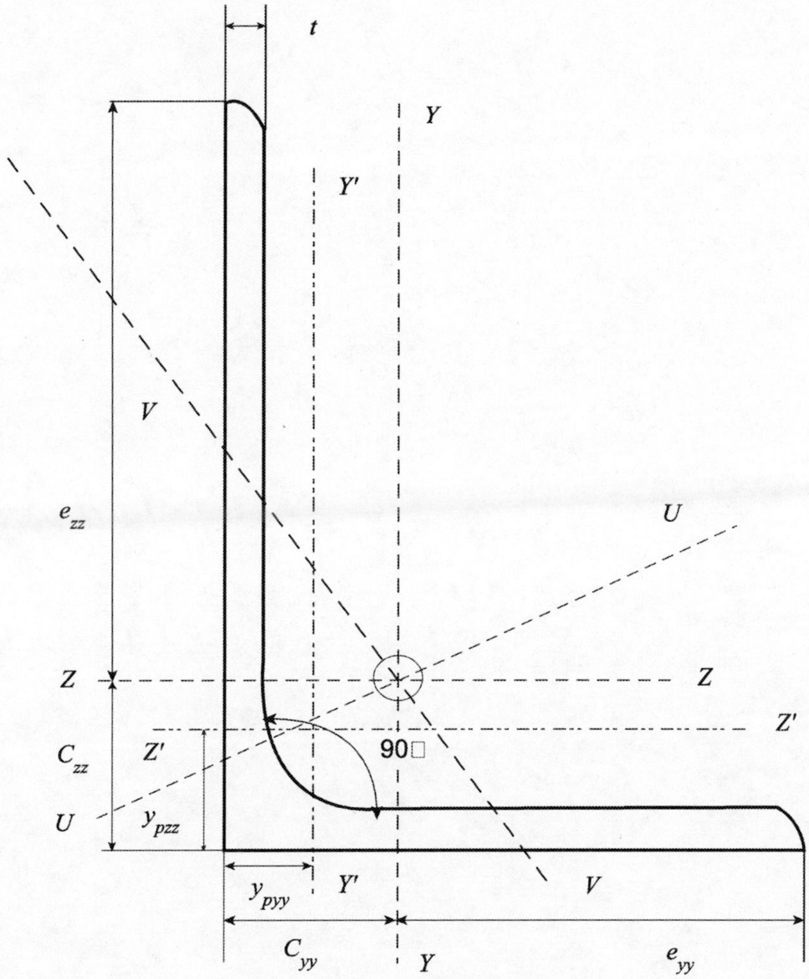

Figure 6 Unequal Angles

Table IV Rolled Steel Unequal Angles

Designation	Thickness	Sectional Area	Weight per metre	Centre of Gravity		Distance of Extreme Fibre		Moments of Inertia				Radii of Gyration				Distance of Plastic N-A		Plastic Moduli of Section		tan α	Radius at Root	Product of inertia	Section Classification		Designation
	t	A	w	C_{zz}	C_{yy}	e_{zz}	e_{yy}	I_{xx}	I_{yy}	I_{uu}	I_{vv}	r_{zz}	r_{yy}	r_{uu}	r_{vv}	Yp_{zz}	Yp_{yy}	Zp_{zz}	Zp_{yy}		r_1	I_{zy}	Bending Comp.	Axial Comp.	
	mm	mm^2	N	mm	mm	mm	mm	cm^4	cm^4	cm^4	cm^4	mm	mm	mm	mm	mm	mm	mm^3	mm^3		mm	cm^4			
ISA 3020	3.0	141.48	10.9	9.8	4.9	20.2	15.1	1.2	0.4	1.4	0.2	9.2	5.4	9.9	4.1	5.8	2.4	1103.87	529.53		4.5	0.4			ISA 3020
	4.0	184.48	14.2	10.2	5.3	19.8	14.7	1.5	0.5	1.8	0.3	9.2	5.4	9.8	4.1	6.5	3.0	1433.27	703.01		4.5	0.5			
	5.0	225.48	17.4	10.6	5.7	19.4	14.3	1.9	0.6	2.1	0.4	9.1	5.3	9.7	4.1	7.1	3.7	1738.52	871.69		4.5	0.6			
ISA 4025	3.0	187.50	14.4	13.0	5.7	27.0	19.3	3.0	0.9	3.3	0.5	12.5	6.8	13.3	5.2	8.1	2.3	2010.34	854.24		5.0	0.9			ISA 4025
	4.0	245.50	18.9	13.5	6.2	26.5	18.8	3.8	1.1	4.3	0.7	12.5	6.8	13.2	5.2	8.8	3.0	2623.19	1135.96		5.0	1.2			
	5.0	301.50	23.2	13.9	6.6	26.1	18.4	4.6	1.4	5.1	0.8	12.4	6.7	13.1	5.2	9.5	3.7	3201.79	1411.67		5.0	1.4			
	6.0	355.50	27.4	14.3	6.9	25.7	18.1	5.4	1.6	5.9	1.0	12.3	6.6	12.9	5.2	10.1	4.4	3747.78	1682.56		5.0	1.6			
ISA 4530	3.0	217.50	16.7	14.2	6.9	30.8	23.1	4.4	1.5	5.0	0.9	14.2	8.4	15.2	6.3	8.1	2.4	2615.63	1236.21		5.0	1.5			ISA 4530
	4.0	285.50	22.0	14.7	7.3	30.3	22.7	5.7	2.0	6.5	1.1	14.1	8.4	15.1	6.3	8.8	3.2	3423.53	1639.74		5.0	1.9			
	5.0	351.50	27.1	15.1	7.7	29.9	22.3	6.9	2.4	7.9	1.4	14.0	3.3	15.0	6.3	9.5	3.9	4192.14	2033.23		5.0	2.3			
	6.0	415.50	32.0	15.5	8.1	29.5	21.9	8.0	2.8	9.2	1.7	13.9	8.2	14.9	6.3	10.1	4.6	4923.13	2418.05		5.0	2.7			
ISA 5030	3.0	233.63	18.0	16.3	6.5	33.7	23.5	5.9	1.6	6.5	1.0	15.9	8.2	16.7	6.5	10.4	2.3	3184.67	1254.29		5.5	1.7			ISA 5030
	4.0	306.63	23.6	16.8	7.0	33.3	23.0	7.7	2.1	8.5	1.2	15.8	3.2	16.6	6.3	11.2	3.0	4168.48	1669.37		5.5	2.3			
	5.0	377.63	29	17.2	07.4	32.8	2.26	9.3	2.5	10.3	1.5	15.7	8.1	16.5	6.3	11.85	3.74	5107.94	2077.19	0.35		2.7	Compact	Semi Comp.	
	6.0	446.63	34	17.6	07.8	32.4	2.22	10.9	2.9	11.9	1.8	15.6	8	16.4	06.3	12.46	4.41	6004.73	2479.00	0.35		3.1	Plastic	Semi Comp.	
ISA 6040	5.0	475.86	36	19.5	09.6	40.5	3.04	16.9	6.0	19.5	3.4	18.9	11.2	20.2	08.5	11.73	3.95	7637.00	3644.44	0.44	6.00	5.8	Semi Comp.	Semi Comp.	ISA 6040
	6.0	564.86	43	19.9	10.0	40.1	3.00	19.9	7.0	22.8	40	18.8	11.1	20.1	08.5	12.36	4.68	9029.82	4349.84	0.43		6.8	Compact	Semi Comp.	
	8.0	736.86	57	20.7	10.8	39.3	2.92	25.4	8.0	29.0	5.2	18.6	11.0	19.8	08.1	13.52	6.07	11663.59	5724.58	0.42		8.5	Plastic	Semi Comp.	
ISA 6545	5.0	525.86	40	20.7	10.8	44.3	3.42	22.1	8.6	25.9	4.8	20.5	12.8	22.2	9.6	11.73	4.04	9119.84	4631.59	0.47	6.00	8.0	Semi Comp.	Semi Comp.	ISA 6545
	6.0	624.86	48	21.1	11.2	43.9	3.38	26.0	10.1	30.4	5.7	20.4	12.7	22.1	9.5	12.36	4.78	10797.65	5524.60	0.47		9.4	Semi Comp.	Semi Comp.	
	8.0	816.86	63	21.9	12.0	43.1	3.3	33.2	12.8	8.7	7.4	20.2	12.5	21.8	09.5	13.52	6.23	13985.42	7262.26	0.46		11.8	Plastic	Semi Comp.	
ISA 7045	5.0	552.20	42	22.7	10.4	47.3	3.46	27.2	8.8	30.9	5.1	22.2	12.6	23.6	09.6	14.09	3.93	10475.21	4678.91	0.41	6.50	8.9	Semi Comp.	Semi Comp.	ISA 7045
	6.0	656.20	51	23.2	10.9	46.8	3.41	32.0	10.3	36.3	6.0	22.1	12.5	23.5	9.6	14.74	4.66	12406.29	5590.22	0.41		10.5	Semi Comp.	Semi Comp.	
	8.0	858.20	66	24.0	11.6	46	3.34	41.0	13.1	16.3	7.8	21.9	12.4	23.2	9.5	15.93	6.07	16086.45	7372.86	0.40		13.2	Plastic	Semi Comp.	
	10.0	1052.20	81	24.8	12.4	45.2	32.6	49.3	15.6	55.4	9.5	21.6	12.2	22.9	9.5	17.05	7.42	19534.26	9110.82	0.39		15.5	Plastic	Semi Comp.	
ISA 7550	5.0	602.20	46	23.9	11.6	51.1	38.4	34.1	12.2	39.4	69	23.8	14.2	25.6	10.7	14.09	4.01	12208.04	5791.42	0.44	6.50	11.8	Semi Comp.	Semi Comp.	ISA 7550
	6.0	716.20	57	24.4	12	50.6	38.0	40.3	14.3	46.4	8.2	23.7	14.1	25.5	10.7	14.74	4.75	14474.13	6915.43	0.44		13.9	Semi Comp.	Semi Comp.	
	8.0	938.20	73	25.2	12.8	49.8	37.2	51.8	18.3	59.4	10.6	23.5	14.0	25.2	10.6	15.93	6.20	18809.28	9111.09	0.43		17.7	Plastic	Semi Comp.	
	10.0	1152.20	88	26	13.6	49.0	36.4	62.3	21.8	71.2	12.9	23.3	13.8	24.9	10.6	17.05	7.60	22892.10	11246.34	0.42		20.9	Plastic	Semi Comp.	
ISA 8050	5.0	626.82	47	26	11.2	54	38.8	40.6	12.2	45.7	7.2	25.5	14	27	10.7	16.45	3.91	13689.80	5797.19	0.39	7.00	12.9	Slender	Semi Comp.	ISA 8050
	6.0	745.82	58	26.4	11.6	53.6	38.4	48.0	14.4	53.9	8.5	25.4	13.9	26.9	10.7	17.12	4.64	16247.56	6940.16	0.39		15.2	Semi Comp.	Semi Comp.	
	8.0	977.82	76	27.3	12.4	52.7	37.6	61.9	18.5	69.3	11.0	25.2	13.7	26.6	10.6	18.34	6.06	21150.88	9182.46	0.38		19.3	Semi Comp.	Semi Comp.	

Contd...

Designation	Thickness	Area	Weight per metre	Centre of Gravity		Distance of Extreme Fibre		Moments of Inertia				Radii of Gyration				Distance of Plastic N-A		Plastic Moduli of Section		tan α	Radius at Root	Product of inertia	Section Classification		Designation
	t	A	w	C_{zz}	C_{yy}	e_{zz}	e_{yy}	I_{xx}	I_{yy}	I_{uu}	I_{vv}	r_{zz}	r_{yy}	r_{uu}	r_{vv}	Yp_{zz}	Yp_{yy}	Zp_{z}	Zp_{y}		r_1	I_{zy}	Bending Comp.	Axial Comp.	
	mm	mm²	N	mm	mm	mm	mm	cm⁴	cm⁴	cm⁴	cm⁴	mm	mm	mm	mm	mm	mm	mm³	mm³		mm	cm⁴			
	10.0	1201.82	92	29.1	13.2	51.9	36.8	74.7	22.1	83.3	13.5	24.9	13.6	26.3	10.6	19.47	7.42	25781.75	11375.37	0.38		22.9	Compact	Semi Comp.	
ISA 9060	6.0	865.34	67	28.7	13.9	61.3	46.1	70.6	25.2	81.5	14.3	28.6	17.1	30.7	12.8	16.99	4.80	21051.90	9930.87	0.44	7.50	24.50	Plastic	Semi Comp.	ISA 9060
	8.0	1137.34	87	29.6	15.5	60.4	45.2	91.5	32.4	105.3	18.6	28.4	16.9	30.4	12.8	18.25	6.28	27516.07	13200.86	0.44		31.50	Semi Comp.	Semi Comp.	
	10.0	1401.34	108	30.4	15.5	59.6	44.5	110.9	39.1	127.3	22.8	28.1	16.7	30.1	12.7	19.40	7.72	33667.64	16339.45	0.43		37.80	Semi Comp.	Semi Comp.	
	12.0	1657.34	128	31.2	16.3	58.8	43.7	129.1	45.2	147.5	26.8	27.9	16.5	29.8	12.7	20.50	9.11	39518.91	19407.80	0.42		43.30	Plastic	Semi Comp.	
ISA 10065	6.0	954.75	74	31.9	14.7	68.1	50.3	96.7	32.4	110.6	18.6	31.8	18.4	34.0	13.9	19.36	4.77	25973.03	11746.60	0.42	8.00	32.50	Plastic	Semi Comp.	ISA 10065
	8.0	1256.75	97	32.8	15.5	67.2	49.3	125.9	41.9	143.6	24.2	31.6	18.3	33.8	13.9	20.64	6.25	34035.74	15576.74	0.42		42.00	Slender	Slender	
	10.0	1550.75	120	33.7	16.3	66.3	48.7	153.2	50.7	174.2	29.7	31.4	18.1	33.5	13.8	21.81	7.69	41745.67	19320.49	0.41		50.70	Semi Comp.	Semi Comp.	
ISA 10075	6.0	1014.05	78	30.1	17.8	69.9	57.2	100.9	48.7	124.0	25.6	31.5	21.9	35.0	15.9	14.21	5.09	26693.13	15502.06	0.55	8.50	41.00	Compact	Slender	ISA 10075
	8.0	1336.05	103	31.0	18.7	69.0	56.3	131.6	63.3	161.3	33.6	31.4	21.8	34.8	15.8	15.53	6.68	35062.08	20509.66	0.55		53.4	Slender	Semi Comp.	
	10.0	1650.05	128	31.9	19.5	68.1	55.5	160.4	76.9	196.1	41.2	31.2	21.6	34.5	15.8	16.72	8.22	43073.03	25372.93	0.55		64.7	Semi Comp.	Semi Comp.	
	12.0	1956.05	151	32.7	20.3	67.3	54.7	187.5	89.5	228.4	41.2	31.0	21.4	34.2	15.8	17.85	9.72	50743.48	30105.65	0.54		74.9	Compact	Semi Comp.	
ISA 12575	6.0	1165.93	90	40.5	15.9	84.5	59.1	187.8	51.6	208.9	30.5	40.1	21.0	42.3	16.2	26.55	4.66	40344.53	15827.71	0.37	9.00	56.7	Slender	Slender	ISA 12575
	8.0	1537.93	119	41.5	16.8	83.5	58.2	245.5	67.2	272.8	40.0	40	20.9	42.1	16.1	27.91	6.12	53044.75	21071.52	0.36		74	Semi Comp.	Semi Comp.	
	10.0	1901.93	146	42.4	17.6	82.6	57.4	300.3	81.6	332.9	49.1	39.7	20.7	41.8	16.1	29.13	7.55	65291.70	26230.37	0.36		89.9	Semi Comp.	Semi Comp.	
ISA 12595	6.0	1285.93	99	37.0	22.2	88.0	72.8	203.2	102.1	252.3	52.9	39.7	28.2	44.3	20.3	16.55	5.17	42570.70	25276.46	0.57	9.00	84.5	Slender	Slender	ISA 12595
	8.0	1697.93	131	38.0	23.1	87.0	71.9	266.0	133	329.7	69.6	39.6	28.0	44.1	20.2	17.91	6.80	56070.92	33469.93	0.57		110.6	Semi Comp.	Semi Comp.	
	10.0	2101.93	162	38.8	23.9	86.2	71.1	325.8	162.7	402.9	85.6	39.4	27.8	43.8	20.2	19.13	8.39	69117.88	41459.45	0.57		135	Semi Comp.	Semi Comp.	
	12.0	2497.93	192	39.6	24.7	85.4	70.3	382.6	190.4	472.0	101.0	39.1	27.6	43.5	20.1	20.28	9.96	81724.20	49259.95	0.56		157.7	Compact	Semi Comp.	
ISA 15075	8.0	1742.01	135	52.3	15.3	97.7	59.7	407.2	70.2	432.8	44.5	48.3	20.1	49.8	16	40.16	5.77	73637.54	21593.93	0.27	10.00	95.5	Slender	Slender	ISA 15075
	10.0	2156.01	166	53.2	16.1	96.8	58.9	499.1	85.3	529.8	54.6	48.1	19.9	49.6	15.9	41.43	7.12	90743.91	27025.88	0.26		116.2	Semi Comp.	Semi Comp.	
	12.0	2562.01	197	54.1	16.9	95.9	58.1	587.0	99.5	622.2	64.3	47.9	19.7	49.3	15.8	42.61	8.45	107309.32	32433.06	0.26		135.2	Semi Comp.	Semi Comp.	
ISA 150115	8.0	2057.82	159	44.6	27.3	105.4	87.7	465.7	238.9	581.2	123.3	47.6	34.1	53.1	24.5	19.88	6.88	81429.55	49189.40	0.58	11.00	195.9	Slender	Slender	ISA 150115
	10.0	2551.82	196	45.5	28.2	105.5	86.8	573.3	293.4	714.3	152.4	47.4	33.9	52.9	24.4	21.20	8.51	100741.44	61112.54	0.58		241	Semi Comp.	Semi Comp.	
	12.0	3037.82	233	46.4	29.0	103.6	86.0	676.5	345.3	841.4	180.4	47.2	33.7	52.6	24.4	22.42	10.11	119511.92	72787.95	0.58		283.6	Semi Comp.	Semi Comp.	
	15.0	3751.82	289	47.6	30.2	102.4	84.8	823.5	418.6	1020.9	221.2	46.9	33.4	52.2	24.3	24.13	12.46	146680.00	89869.23	0.57		342.8	Compact	Semi Comp.	
ISA 200100	10.0	2903.43	224	69.6	20.1	130.4	79.9	1210	209.2	1286.7	132.5	64.6	26.8	66.6	21.4	53.45	7.22	163924.36	47906.60	0.27	12.00	284.8	Slender	Slender	ISA 200100
	12.0	3459.43	267	70.5	21.0	129.5	79.0	1431.7	246.2	1521.0	156.8	64.3	26.7	66.3	21.3	54.71	8.58	194706.90	57575.90	0.26		335.3	Slender	Semi Comp.	
	15.0	4278.43	330	71.8	22.2	128.2	77.8	1750.5	298.1	1856.7	191.9	64.0	26.4	65.9	21.2	56.47	10.59	239517.48	72021.90	0.26		405.4	Semi Comp.	Semi Comp.	
ISA 200150	10.0	3400.38	262	59.9	35.1	140.1	114.9	1377.9	669.6	1696.6	350.8	63.7	44.4	70.6	32.1	28.04	8.53	180946.16	104993.52	0.56	13.50	564.1	Slender	Slender	ISA 200150
	12.0	4056.38	312	60.8	36.0	139.2	114	1634.9	793.2	2010.8	417.2	63.5	44.2	70.4	32.1	29.37	10.15	215484.16	125469.55	0.56		669.1	Slender	Semi Comp.	
	15.0	5025.38	387	62.0	37.2	138.0	112.8	2005.6	969.9	2461.9	513.6	63.2	43.9	70.0	32.0	31.20	12.54	265926.72	155598.40	0.55		818.5	Semi Comp.	Semi Comp.	
	18.0	5976.38	460	63.3	38.4	136.7	111.6	2359.4	1136.9	2889.5	606.9	62.8	43.6	69.5	31.9	32.91	14.89	314769.65	185037.49			958.1	Semi Comp.	Semi Comp.	

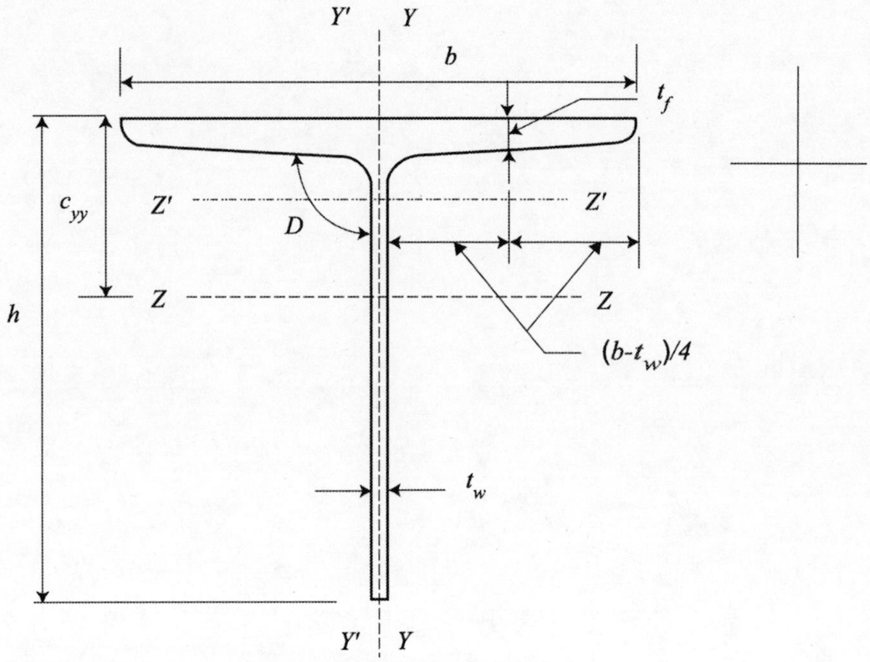

Figure 7 T-Sections

Designation	Weight per metre	Sectional Area	Depth of Section	Width of Flange	Thickness of Flange	Thickness of Web	Centre of Gravity	Moments of Inertia		Radii of Gyration		Moduli of section		Radius at Root	Radius at Toe	Flange	Section Classification		Plastic N-A	Plastic Modulus of Sections	
	w	a	h	b	t_f	t_u	C_y	I_{zz}	I_{yy}	r_z	r_y	Z_{zz}	Z_{yy}	r_1	r_2		Stem	Entire Section	yp_z	ZP_{zz}	ZP_{yy}
	N	mm^2	mm	mm	mm	mm	mm	cm^4	cm^4	mm	mm	cm^3	cm^3	mm	mm				mm	mm^3	mm^3
ISNT 20	9	113.88	20	20	3.0	3.0	6	0.4	0.2	5.9	3.9	0.3	0.2	4.0	3.0	Plastic	Plastic	Plastic	2.78	526.81	317.68
ISNT 30	14	176.25	30	30	3.0	3.0	8	1.4	0.6	8.9	5.7	0.6	0.4	5.0	3.5	Plastic	Semi compact	Semi compact	2.85	1235.43	683.86
ISNT 40	34	449.87	40	40	6.0	6.0	12	6.3	3.0	11.8	8.2	2.2	1.5	5.5	4.0	Plastic	Plastic	Plastic	5.55	4200.22	2618.55
ISNT 50	44	572.24	50	50	6.0	6.0	14	12.7	5.9	15.0	10.2	3.6	2.4	6.0	4.0	Plastic	Plastic	Plastic	5.64	6725.44	4020.60
ISNT 60	53	693.07	60	60	6.0	6.0	17	22.5	10.1	18.1	12.1	5.2	3.4	6.5	4.6	Plastic	Semi compact	Semi compact	5.70	9848.21	5664.47
ISNT 80	94	1229.91	80	80	8.0	8.0	22	71.2	32.3	24.1	16.2	12.3	8.1	8.0	5.5	Plastic	Semi compact	Semi compact	7.60	23337.69	13485.92
ISNT 100	147	1918.54	100	100	10.0	10.0	28	79.9	173.8	30.2	20.5	24.1	16.0	9.0	6.0	Plastic	Semi compact	Semi compact	9.50	45571.10	26497.49
ISNT 150	224	2921.01	150	150	10.0	10.0	40	267.5	603.8	45.6	30.3	54.6	35.7	10.0	7.0	Plastic	Semi compact	Semi compact	9.67	105588.11	57568.11
ISHT 75	150	1948.87	75	150	9.0	8.4	16	230.2	96.2	22.2	34.4	16.4	30.1	8.0	4.0	Semi compact	Compact	Compact	6.31	21927.83	49125.28
ISHT 100	196	2547.22	100	200	9.0	7.8	19	497.3	193.8	27.6	44.2	24.0	49.3	9.0	4.5	Semi compact	Semi compact	Semi compact	6.26	37044.31	81733.30
ISHT 125	267	3485.37	125	250	9.7	8.8	24	1005.8	415.4	34.5	53.7	41.0	79.9	10.0	5.0	Semi compact	Semi compact	Semi compact	6.86	65496.77	134275.92
ISHT 150	288	3742.48	150	250	10.6	7.6	27	573.7	1096.8	39.2	54.1	46.5	87.7	11.0	5.5	Plastic	Slender	Slender	7.42	82058.97	145791.76
ISST 100	81	1036.76	100	50	10.0	5.8	30	99.0	9.6	30.9	09.6	14.2	3.8	8.0	4.0	Plastic	Semi compact	Semi compact	11.90	25960.25	6720.52
ISST 150	154	1995.87	150	75	11.6	8.0	48	450.2	37.0	47.5	13.6	43.9	9.9	9.0	4.5	Plastic	Semi compact	Semi compact	26.43	83366.55	17558.52
ISST 200	279	3621.52	200	165	12.5	8.0	48	1267.8	358.2	59.2	31.5	83.3	43.4	16.0	8.0	Plastic	Slender	Slender	10.80	150752.96	75697.88
ISST 250	368	4774.91	250	180	14.1	9.2	64	2774.4	532.0	76.2	33.4	149.2	59.1	17.0	8.5	Plastic	Slender	Slender	13.08	271880.65	103332.92
ISLT 50	39	510.55	50	50	6.4	4.0	12	9.9	6.4	13.9	11.2	2.6	2.5	7.0	3.0	Plastic	Semi compact	Semi compact	4.94	4495.46	4103.08
ISLT 75	170	904.16	75	80	6.8	4.8	17	41.9	27.6	21.5	17.5	7.2	6.9	9.5	3.0	Plastic	Semi compact	Semi compact	5.45	12493.14	11061.94
ISLT 100	124	1616.34	100	100	10.8	5.7	21	116.6	75.0	26.9	21.5	14.8	15.0	11.0	5.5	Plastic	Semi compact	Semi compact	7.94	26311.57	24994.45
ISJT 75	34	450.38	75	50	4.6	3.0	20	24.8	4.6	23.5	10.1	4.5	1.8	5.0	1.5	Plastic	Slender	Slender	4.41	7978.14	2980.66
ISJT 87.5	39	513.82	87.5	50	4.8	3.2	25	39.0	4.8	27.5	09.7	6.2	1.9	5.0	1.5	Plastic	Slender	Slender	8.65	11520.46	3160.70
ISJT 100	49	632.18	100	60	5.0	3.4	26	63.5	8.6	31.7	11.7	8.8	2.9	5.0	1.5	Plastic	Slender	Slender	8.38	16092.61	4677.01
ISJT 112.5	63	813.90	112.5	80	5.0	3.7	30	101.6	20.2	35.3	15.8	12.3	5.1	6.5	1.5	Plastic	Slender	Slender	4.99	22399.62	8145.64

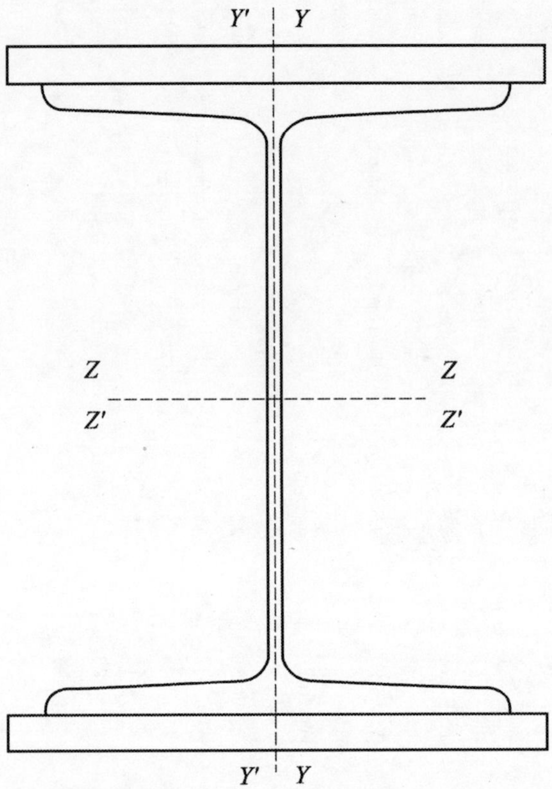

Figure 8 I-Sections With Cover Plates on Both Flanges to be Used as Girders

Section Classification: $t_f \leq 40$ mm about z-z axis *b* about y-y axis *c* $t_f > 40$ mm about z-z axis *c* about y-y axis *d*

Composed of														
Joist	Cover Plates		Weight per metre	Sectional Area	Mean Thickness of Flanges	Extreme Fibre Distance	Gross Moments of Inertia		Least Radius of Gyration	Modulus of Section	Plastic Modulus of Sections		
Designation	Width	Thickness	w	A		e_y	I_{zz}	I_{yy}	r_{yy}	Z_{zz}	Zp_{zz}	Zp_{yy}	Designation
	mm	mm	N	mm^2	mm	mm	cm^4	cm^4	mm	cm^3	mm^3	mm^3	
ISWB 150	160.0	10.0	413.0	5396.53	14.4	85.0	2889.8	777.5	38.1	340.0	382,857.02	159,938.83	ISWB 150
		12.0	463.0	6006.53	16.4	87.0	363.1	914.0	39.0	386.6	437,897.02	185,538.83	
		16.0	561.1	7286.53	20.4	91.0	4377.2	1187.1	40.4	481.0	551,817.02	236,738.83	
		20.0	660.2	8596.53	24.4	95.0	5484.4	1460.1	41.3	577.3	670,857.02	287,938.83	
		25.0	782.8	10166.53	29.4	100.0	7005.8	1801.5	42.1	700.6	826,857.02	351,938.83	
ISWB 175	200.0	10.0	522.9	6811.29	14.6	97.5	4935.2	1521.9	47.3	506.2	564,189.53	251,272.20	ISWB 175
		12.0	585.7	7611.29	16.6	99.5	5711.3	1788.6	48.5	574.0	642,989.53	291,272.20	
		16.0	709.3	9211.29	20.6	103.2	7360.0	2321.9	50.2	711.1	805,389.53	371,272.20	
		20.0	832.9	10811.29	24.6	107.5	9141.1	2855.3	41.4	850.3	974,189.53	451,272.20	
		25.0	986.9	12811.29	29.6	112.5	11561.5	3521.9	52.4	1027.7	1,194,189.53	551,272.20	
ISMB 200	160.0	10.0	495.4	6432.67	16.8	110.0	5766.1	832.7	36.0	524.2	589,855.61	177,988.90	ISMB 200
		12.0	544.5	7072.67	18.8	112.0	6554.6	969.2	37.0	585.2	660,895.61	203,588.90	
		16.0	643.5	8352.67	22.8	116.0	8218.3	1242.3	38.6	708.5	806,815.61	254,788.90	
		20.0	741.6	9632.67	26.8	120.0	10000.7	1515.3	39.7	833.4	957,855.61	305,988.90	
		25.0	865.2	11232.67	31.8	125.0	12402.1	1856.7	40.4	992.2	1,153,855.61	369,988.90	
ISWB 200	200.0	10.0	590.6	7670.87	16.3	110.0	7037.8	1662.1	46.6	639.8	713,984.76	278,702.39	ISWB 200
		12.0	652.4	8470.87	18.3	112.0	8023.5	1928.8	47.7	716.4	802,784.76	318,702.39	
		16.0	776.0	10070.87	22.3	116.0	10103.1	1262.1	49.4	871.0	985,184.76	398,702.39	
		20.0	898.6	11670.87	26.3	120.0	12331.2	2995.5	50.7	1027.6	1,173,984.76	478,702.39	
		25.0	1052.6	13670.87	31.3	125.0	15332.8	3662.1	51.8	1226.6	1,418,984.76	578,702.39	
ISMB 225	160.0	10.0	552.3	7171.50	18.1	122.5	7862.4	901.0	35.4	641.8	724,270.85	194,313.60	ISMB 225
		12.0	601.4	7811.50	20.1	124.5	8838.5	1037.5	36.4	709.6	803,310.85	219,913.60	
		16.0	700.4	9091.50	24.1	128.5	10887.0	1310.6	38.0	847.2	965,230.85	271,113.60	
		20.0	798.5	10371.50	28.1	132.5	13067.0	1583.6	39.1	986.2	1,132,270.85	322,313.60	
		25.0	922.1	11971.50	33.1	137.5	15983.5	1925.0	40.1	1162.4	1,348,270.85	386,313.60	
ISWB 225	200.0	10.0	640.6	8323.95	17.4	122.5	1781.9	9446.2	46.3	771.1	859,930.93	299,771.54	ISWB 225
		12.0	702.4	9123.95	19.4	124.5	10666.4	2048.6	47.4	856.7	958,730.93	339,771.54	
		16.0	826.0	10723.95	23.4	128.5	13227.0	2581.9	49.1	1029.3	1,161,130.93	419,771.54	
		20.0	948.6	12323.95	27.4	132.5	15952.0	3115.3	50.3	1203.9	1,369,930.93	499,771.54	
		25.0	1102.6	14323.95	32.4	137.5	19597.6	3781.9	51.4	1425.3	1,639,930.93	599,771.54	

Contd...

Composed of													
Joist	Cover Plates		Weight per metre	Sectional Area	Mean Thickness of Flanges	Extreme Fibre Distance	Gross Moments of Inertia		Least Radius of Gyration	Modulus of Section	Plastic Modulus of Sections		Designation
Designation	Width	Thickness	w	A		e_{zz}	I_{zz}	I_{yy}	r_{yy}	Z_{zz}	Zp_{zz}	Zp_{yy}	
	mm	mm	N	mm^2	mm	mm	cm^4	cm^4	mm	cm^3	mm^3	mm^3	
ISMB 250	200.0	10.0	673.9	8755.43	17.8	135.0	11894.9	1667.8	43.6	881.1	985,712.95	289,701.12	ISMB 250
		12.0	735.8	9555.43	19.8	137.0	13374.6	1934.5	45.0	976.3	1,094,512.95	329,701.12	
		16.0	859.4	11155.43	23.8	141.0	16466.2	2467.8	47.0	1167.8	1,316,912.95	409,701.12	
		20.0	982.0	12755.43	27.8	145.0	19738.3	3001.2	48.5	1361.1	1,545,712.95	489,701.12	
		25.0	1136.0	14755.43	32.3	150.0	24089.9	3667.8	49.9	1606.0	1,840,712.95	589,701.12	
ISWB 250	320.0	10.0	893.7	11604.61	15.6	135.0	16764.4	6318.8	73.8	1241.8	1359556.03	661669.92	ISWB 250
		12.0	991.8	12884.61	17.6	137.0	19132.0	7411.1	75.8	1396.5	1533636.03	764069.92	
		16.0	1189.0	15444.61	21.6	141.0	24078.5	9595.6	78.8	1707.7	1889476.03	968869.92	
		20.0	1386.2	18004.61	25.6	145.0	29313.8	11780.2	80.9	2021.6	2255556.03	1173669.92	
		25.0	1633.4	21204.61	30.6	150.0	36276.0	14510.8	82.7	2418.4	2727556.03	1429669.92	
ISMB 300	200.0	10.0	741.6	9626.38	18.7	160.0	18216.9	1787.2	43.1	1138.6	1271730.75	310457.98	ISMB 300
		12.0	802.5	10426.38	20.7	162.0	20290.6	2053.9	44.4	1252.5	1400530.75	350457.98	
		16.0	926.1	12026.38	24.7	166.0	24594.2	2587.2	46.4	1481.6	1662930.75	430457.98	
		20.0	1049.7	13626.38	28.7	170.0	29110.3	3120.6	47.9	1712.4	1931730.75	510457.98	
		25.0	1203.7	15626.38	33.7	175.0	35061.9	3787.2	49.2	2003.5	2276730.75	610457.98	
		32.0	1418.5	18426.38	40.7	182.0	43984.5	4720.6	50.6	2416.7	2776530.75	750457.98	
ISWB 300	320.0	10.0	965.3	12532.75	16.2	160.0	25202.9	6451.4	71.7	1575.2	1723207.54	683014.22	ISWB 300
		12.0	1063.4	13812.75	18.2	162.0	28520.9	7543.7	43.9	1760.5	1929287.54	785414.22	
		16.0	1260.6	16372.75	22.2	166.0	35406.6	9728.2	77.1	2132.9	2349127.54	990214.22	
		20.0	1457.8	18932.75	26.2	170.0	42632.3	11912.8	79.3	2507.8	2779207.54	1195014.22	
		25.0	1704.0	22132.75	31.2	175.0	52154.5	14643.4	81.3	2980.3	3331207.54	1451014.22	
		32.0	2059.1	26612.75	38.2	182.0	66431.1	18466.4	83.3	3650.1	4130887.54	1809414.22	
ISMB 350	200.0	10.0	822.1	10671.34	19.9	185.0	26593.6	1871.0	41.9	1437.5	1609576.00	329723.39	ISMB 350
		12.0	882.9	11471.34	21.9	187.0	29361.3	2137.0	43.2	1570.1	1758376.00	369723.39	
		16.0	1006.5	13071.34	25.9	191.0	35076.9	2671.0	45.2	1836.5	2060776.00	449723.39	
		20.0	1130.1	14671.34	29.9	195.0	41037.0	3204.4	46.7	2104.5	2369576.00	529723.39	
		25.0	1284.1	16671.34	34.9	200.0	48838.6	3871.0	48.2	2441.9	2764576.00	629723.39	
		32.0	1499.0	19471.34	41.9	207.0	60435.2	4804.4	49.7	2919.6	3334376.00	769723.39	
ISWB 350	320.0	10.0	1051.6	13649.90	17.1	135.0	36263.0	6637.2	69.7	1960.2	2147484.03	712466.66	ISWB 350
		12.0	1149~7	14929.90	19.1	187.0	40691.4	7729.5	72.0	2176	2385564.03	814866.66	
		16.0	1346.9	17489.90	23.1	191.0	49836.3	9914.0	75.3	2609.2	2869404.03	1019666.66	
		20.0	1544.1	20049.90	27.1	195.0	59372.4	12098.6	77.7	3044.7	3363484.03	1224466.66	
		25.0	1790.3	23249.90	32.1	200.0	71855.0	14829.2	79.9	3592.8	3995484.03	1480466.66	

Contd...

Composed of

Joist	Cover Plates		Weight per metre	Sectional Area	Mean Thickness of Flanges	Extreme Fibre Distance	Gross Moments of Inertia		Least Radius of Gyration	Modulus of Section	Plastic Modulus of Sections		
Designation	Width	Thickness	w	A		e_{zz}	I_{zz}	I_{yy}	r_{yy}	Z_{zz}	Zp_{zz}	Zp_{yy}	Designation
	mm	mm	N	mm^2	mm	mm	cm^4	cm^4	mm	cm^3	mm^3	mm^3	
		32.0	2135.6	27729.90	39.1	207.0	90409.6	18652.2	82.0	4367.6	4907164.03	1838866.66	
ISMB 400	200.0	10.0	912.3	11845.58	21.2	210.0	37271.7	1955.4	40.6	1774.8	1996163.26	349666.89	ISMB 400
		12.0	974.1	12645.58	23.2	212.0	40833.4	222.1	41.9	1926.1	2164963.26	389666.89	
		16.0	1091.9	14245.58	27.2	216.0	48161.0	2755.4	44.0	2229.7	2507363.26	469666.89	
		20.0	1220.4	15845.58	31.2	220.0	55756.1	3288.8	45.6	2534.8	2856163.26	549666.89	
		25.0	1374.4	17845.58	36.2	225.0	65666.7	3955.4	47.1	2918.5	3301163.26	649666.89	
		32.0	1590.2	20645.58	43.2	232.0	80287.3	4888.8	48.7	3460.7	3940963.26	789666.89	
ISWB 400	320.0	10.0	98.1	14901.25	18.1	210.0	50328.0	6849.3	67.8	2396.6	2632176.95	746187.39	ISWB 400
		12.0	117.7	16181.25	20.1	212.0	56026.8	7941.6	70.1	2642.8	2902256.95	848587.39	
		16.0	157.0	18741.25	24.1	216.0	67750.9	10126.1	73.5	3136.6	3450096.95	1053387.39	
		20.0	196.2	21301.25	28.1	220.0	79917.4	12310.7	76.0	3632.6	4008176.95	1258187.39	
		25.0	245.3	24501.25	33.1	225.0	95760.0	15041.3	78.4	4256.0	4720176.95	1514187.39	
		32.0	313.92	28981.25	40.1	232	119153	18864.3	80.7	5135.9	5743856.95	1872587.39	
		40.0	392.4	34101.25	48.1	240	147672	23233.3	82.5	6153	6952176.95	2282187.39	
ISMB 450	200.0	10.0	98.1	13226.63	23.0	235.0	51554.1	2167.3	40.5	2193.8	2473347.46	387128.80	ISMB 450
		12.0	117.7	14026.63	25.0	237.0	56009.8	2434.0	41.7	2363.3	2662147.46	427128.80	
		16.0	157.0	15626.63	29.0	241.0	65149.4	2967.3	43.6	2703.3	3044547.46	507128.80	
		20.0	196.2	17226.63	33.0	245.0	74597.5	3500.7	45.1	3044.8	3433347.46	587128.80	
		25.0	245.3	19226.63	38.0	250.0	86849.1	4167.3	46.6	3474.0	3928347.46	687128.80	
		32.0	313.9	22026.63	45.0	257.0	104843.7	5100.7	48.1	4079.5	4638147.46	827128.80	
		40.0	392.4	25226.63	53.0	265.0	126644.1	6167.3	49.4	4779.0	547334'7.46	987128.80	
ISWB 450	320.0	10.0	98.1	16515.05	19.6	235.0	68918.9	7168.0	65.9	2932.7	3232576.17	796175.38	ISWB 450
		12.0	117.7	17795.05	21.6	237.0	76048.1	8260.3	68.1	3208.8	3534656.17	898575.38	
		16.0	157.0	20355.05	25.6	241.0	90671.4	10444.8	71.6	3762.3	4146496.17	1103375.38	
		20.0	196.2	22915.05	29.6	245.0	105788.3	12629.4	74.2	4317.9	4768576.17	1308175.38	
		25.0	245.3	26115.05	34.6	250.0	125390.9	15360.0	76.7	5015.6	5560576.17	1564175.38	
		32.0	313.9	30595.05	41.6	257.0	154182.3	19183.0	79.2	5999.3	6696256.17	1922575.38	
		40.0	392.4	35715.05	49.6	265.0	189062.9	23552.0	81.2	7134.5	8032576.17	2332175.38	
ISMB 500	250.0	10.0	98.1	16074.38	22.4	260.0	77735.0	3974.0	49.7	2989.8	3349650.52	572111.14	ISMB 500
		12.0	117.7	17074.38	24.4	262.0	84547.1	4494.8	51.3	3227.0	3610650.52	634611.14	
		16.0	157.0	19074.38	28.4	266.0	98486.6	5536.5	53.9	3702.5	4138650.52	759611.14	
		20.0	196.2	21074.38	32.4	270.0	112851.6	6578.1	55.9	4179.7	4674650.52	884611.14	
		25.0	245.3	23574.38	37.4	275.0	131416.2	7880.2	57.8	4778.8	5355900.52	1040861.14	

Contd...

Composed of														
Joist	Cover Plates		Weight per metre	Sectional Area	Mean Thickness of Flanges	Extreme Fibre Distance	Gross Moments of Inertia		Least Radius of Gyration	Modulus of Section	Plastic Modulus of Sections		Designation
Designation	Width	Thickness	w	A		e_{zz}	I_{zz}	I_{yy}	r_{yy}	Z_{zz}	Zp_{zz}	Zp_{yy}	
	mm	mm	N	mm^2	mm	mm	cm^4	cm^4	mm	cm^3	mm^3	mm^3	
		32.0	313.9	27074.38	44.4	282.0	158564.4	9703.1	59.9	5622•9	6330650.52	1259611.14	
		40.0	392.4	31074.38	52.4	290.0	191285.0	11786.5	61.6	6596.0	7474650.52	1509611.14	
ISWB 500	400.0	10.0	98.1	20121.91	19.2	260.0	104317.6	13654.5	82.4	4012.2	4393350.98	1206821.59	ISWB 500
		12.0	117.7	21721.91	21.2	262.0	115217.0	15787.8	85.3	4397,6	4810950.98	1366821.59	
		16.0	157.0	24921.91	25.2	266.0	137520.1	20054.5	89.7	5169.9	5655750.98	1686821.59	
		20.0	196.2	28121.91	29.2	270.0	160504.2	24321.1	93.0	5944.6	6513350.98	2006821.59	
		25.0	245.3	32121.91	34.2	275.0	190207.6	29654.5	96.1	6916.6	7603350.98	2406821.59	
		32.0	313.9	37721.91	41.2	282.0	233644.7	37121.1	98.2	8285.3	9162950.98	2966821.59	
		40.0	392.4	44121.91	49.2	290.0	285997.6	45654.5	11.7	9862.0	10993350.98	3606821.59	
ISMB 550	250.0	10.0	1402.8	18211.08	24.7	265.0	104097.8	4438.0	49.4	3652.6	4111958.11	640551.25	ISMB 550
		12.0	1479.3	19211.08	26.7	287.0	112277.4	4958.8	50.8	3912.1	4397958.11	703051.25	
		16.0	1633.4	21211.08	30.7	291.0	128981.9	6000.5	53.2	4432.4	4975958.11	828051.25	
		20.0	1787.4	23211.08	34.7	295.0	146151.9	7042.1	55.1	4954.3	5561958.11	953051.25	
		25.0	1979.7	25711.08	39.7	300.0	168279.0	8344.2	57.0	5609.3	6305708.11	1109301.25	
		32.0	2249.4	29211.08	46.7	307.0	200519.7	10167.1	59.0	6531.6	7367958.11	1328051.25	
		40.0	2557.5	33211.08	54.7	315.0	239210.3	12250.5	60.7	7594	8611958.11	1578051.25	
ISWB 550	400.0	10.0	1719.7	22333.94	21.0	285.0	137632.8	14407.3	80.3	4829.2	530625~.60	1300169.11	ISWB 550
		12.0	1843.3	23933.94	23.0	287.0	150720.2	16540.6	83.1	5251.6	5763852.60	1460169.11	
		16.0	2089.5	27133.94	27.0	291.0	177447.3	20807.3	87.6	6097.8	6688652.60	1780169.11	
		20.0	2335.8	30333.94	31.0	295.0	204919.4	25073.9	90.9	6946.4	7626252.60	2100169.11	
		25.0	2643.8	34333.94	36.0	300.0	240322.8	30407.3	94.1	8010.8	8816252.60	2500169.11	
		32.0	3075.4	39933.94	43.0	307.0	291907.9	37873.9	97.4	9508.4	10515852.60	3060169.11	
		40.0	3567.9	46333.94	51.0	315.0	353812.8	46407.3	100.6	11232.2	12506252.60	3700169.11	
ISMB 600	320.0	10.0	1696.1	22021.24	23.6	310.0	151354.3	8112.3	60.7	4882.4	5462597.89	941318.06	ISMB 600
		12.0	1794.2	23301.24	25.6	312.0	163734.7	9204.6	62.9	5247.9	5860677.89	1043718.06	
		16.0	1991.4	25861.24	29.6	316.0	188975.6	11389.1	66.4	5980.2	6664517.89	1248518.06	
		20.0	2188.6	28421.24	33.6	320.0	214863.7	13573.7	69.1	6714.5	7478597.89	1453318.06	
		25.0	2434.8	31621.24	38.6	325.0	248146.3	16304.3	71.8	7635.3	8510597.89	1709318.06	
		32.0	2780.2	36101.24	45.6	332.0	296492.9	20127.3	74.7	8930.5	9982277.89	2067718.06	
		40.0	3174.5	41221.24	53.6	340.0	354298.3	24496.3	77.1	10420.5	11702597.89	2477318.06	
ISWB 600	400.0	10.0	1927.7	25037.99	23.3	310.0	180625.2	15369.2	78.3	5826.6	6426643.14	1419223.64	ISWB 600
		12.0	2051.3	26637.99	25.3	312.0	196100.6	17502.5	81.1	6285.3	6924243.14	1579223.64	
		16.0	2297.5	29837.99	29.3	316.0	227651.7	21769.2	85.4	7204.2	7929043.14	1899223.64	

Contd...

Joist	Cover Plates		Weight per metre	Sectional Area	Mean Thickness of Flanges	Extreme Fibre Distance	Gross Moments of Inertia		Least Radius of Gyration	Modulus of Section	Plastic Modulus of Sections		
Designation	Width	Thickness	w	A		e_{yy}	I_{zz}	I_{yy}	r_{yy}	Z_{zz}	Zp_{zz}	Zp_{yy}	Designation
	mm	mm	N	mm^2	mm	mm	cm^4	cm^4	mm	cm^3	mm^3	mm^3	
		20.0	2543.7	33037.99	33.3	320.0	260011.8	26035.8	88.8	8125.4	8946643.14	2219223.64	
		25.0	2851.8	37037.99	38.3	325.0	301615.2	31369.2	92	9280.5	10236643.14	2619223.64	
		32.0	3283.4	42637.99	45.3	332.0	362048.3	38835.8	95.4	10905.1	12076243.14	3179223.64	
		40.0	3775.9	49037.99	53.3	340.0	434305.2	47369.2	98.3	12773.7	14226643.14	3819223.64	
ISWB 600	400.0	10.0	2039.5	26485.72	24.8	310.0	190053.3	15965.0	77.6	6130.8	6781606.86	1492908.24	ISWB 600
		12.0	2163.1	28085.72	26.8	312.0	205528.7	18098.3	80.3	6587.5	7279206.86	1652908.24	
		16.0	2409.3	31285.72	30.8	416.0	237079.8	22365	84.5	7502.5	8284006.86	1972908.24	
		20.0	2655.6	34485.72	34.8	320.0	269439.9	26631.6	87.9	8420	9301606.86	2292908.24	
		25.0	2963.6	38485.72	39.8	325.0	311043.3	31965	91.1	9570.6	10591606.86	2692908.24	
		32.0	3395.2	44085.72	46.8	332.0	371476.4	39431.6	94.6	11189	12431206.86	3252908.24	
		40.0	3887.7	50485.72	54.8	340.0	443733.3	47965	97.5	13051	14581606.86	3892908.24	

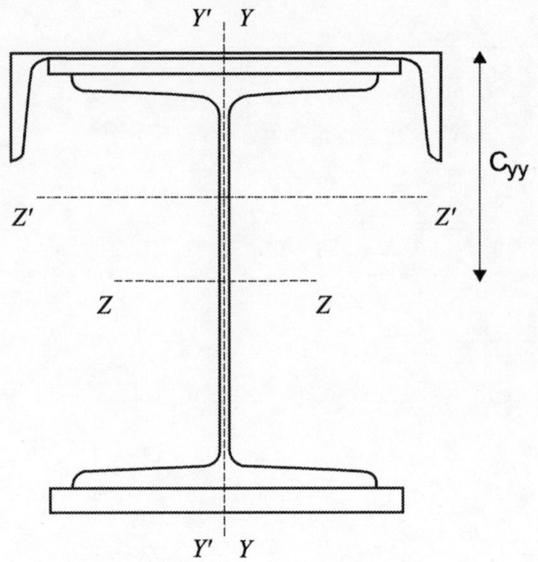

Figure 9 I-Sections With Cover Plates on Flanges and Channel on one Flange to be Used as Girders

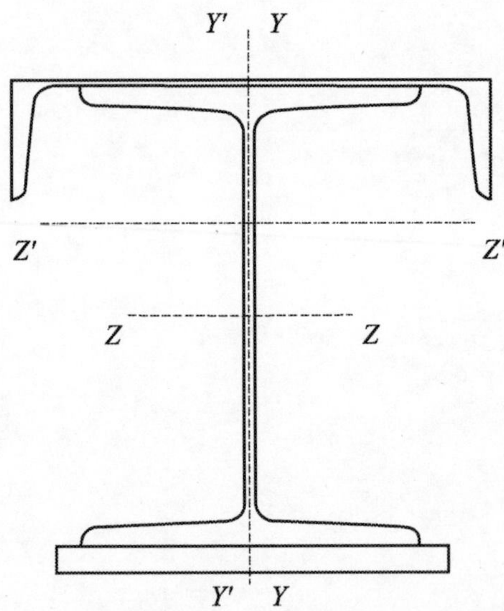

Figure 10 I-Sections With Channel on one Flange and Cover Plate on other Flange to be Used as Girders

Composed of											Mean Thickness of Flanges		Extreme Fibre Distances		Gross Moments of Inertia			Radius of Gyration	Elastic Modulus of Sections		Plastic N-A	Plastic Modulus of Sections	
Joist		Top Flange				Bottom Flange		Weight per Metre	Sectional Area a	Centre of Gravity C_{xx}	Top	Bottom	e_{xx}	e_{yy}	I_{xx}	I_{yy}			Z_{zz}				
Design-ation	W	Channel		Plate		Plate										Whole section	Top Flange Only	r_{yy}	Z_t	Z_l	Yp_{xx}	Zp_{xx}	Zp_{yy}
		Design-ation	w	width x	Thick-ness	width x	Thick-ness																
	kg		N	mm	mm	mm	mm	kg	mm²	cm	mm	mm	cm	cm	cm⁴	cm⁴	cm⁴	cm	cm³	cm³	mm	mm³	mm³
ISMB 600	122.6	ISMC 400	484.6	320 ×	10.0	320 ×	20.0	247.4	31514.57	29.18	27.5	33.6	34.68	20.00	235892.2	25925.8	19138.6	9.07	8.84.1	6801.9	189.71	8124005.86	2088338.22
					12.0		25.0	265.0	33754.57	30.46	29.1	38.6	34.10		260866.4	27837.3	19684.9	9.08	8564.4	7649.9	231.71	8956165.55	2267538.22
					16.0		32.0	292.5	37274.57	31.64	32.3	45.6	33.82		299883.3	30841.0	20777.3	9.10	9418.8	8866.7	275.71	10185231.89	2549138.22
					20.0		40.0	322.7	41114.57	33.34	35.5	53.6	33.52		343302.5	34177.8	21896.7	9.11	10297.7	10241.0	333.04	11506049.24	2856338.22
		ISMC 350	413	250 ×	10.0	320 ×	20.0	234.6	29886.75	30.62	27.7	33.6	33.19	17.50	222893.7	19422.4	12635.4	8.06	7278.4	6716.5	257.04	7788202.98	1769758.51
					12.0		25.0	251.1	31986.75	32.00	29.2	38.6	32.51		245424.1	21048.2	12896.0	8.11	7670.7	7548.1	304.87	8512394.67	1929008.51
					16.0		32.0	276.5	35226.75	33.53	32.0	45.6	32.08		280290.7	23480.5	13417.0	8.16	8358.6	8738.1	360.54	9570547.81	2170708.51
					20.0		40.0	304.5	38786.75	35.17	34.9	53.6	31.64		318880.7	26185.8	13938.0	8.22	9066.2	10079.2	429.54	10662627.69	2438008.51
		ISMC 400	484.6	0	0	320 ×	10.0	197.1	25114.57	27.62	19.5	23.6	34.24	20.00	170351.8	20464.5	16407.8	9.03	6167.7	4975.2	179.71	6077072.60	1576338.22
				0	0		12.0	202.2	25754.57	28.47		25.6	33.59		177711.4	21010.6	16408.0	9.03	6241.4	5291.1	206.38	6350067.95	1627538.22
				0	0		16.0	212.2	27034.57	30.07		29.6	32.39		191631.5	22102.9	16408.2	9.04	6372.2	5916.9	259.71	6848698.67	1729938.22
				0	0		20.0	222.3	28314.57	31.55		33.6	31.31		204609.7	23195.1	16408.4	9.05	6486.1	6534.2	313.04	7284182.72	1832338.22
		ISMC 350	413	0	0	320 ×	10.0	189.9	24186.75	28.55	20.6	23.6	33.26	17.50	163968.8	15389.7	11333.2	7.98	5743.0	4930.1	217.87	5911095.68	1357508.51
				0	0		12.0	194.9	24826.75	29.41		25.6	32.60		170907.2	15935.8	11333.3	8.01	5811.0	5242.7	244.54	6159349.16	1408708.51
				0	0		16.0	204.9	26106.75	31.02		29.6	31.39		184002.1	17028.1	11333.5	8.08	5931.9	5861.7	297.87	6608496.13	1511108.51
				0	0		20.0	215.0	27386.75	32.50		33.6	30.31		196179.3	18120.3	11333.7	8.13	6037.1	6471.5	351.20	6994496.42	1613508.51
		ISMC 300	351.2	0	0	250 ×	10.0	178.1	22684.95	28.41	22.2	27.5	33.35	15.00	150938.0	10315.7	7687.8	6.74	5313.3	4252.5	221.61	5480463.92	1082335.55
				0	0		12.0	182.0	23184.95	29.13		29.5	32.83		156412.7	10576.1	7687.9	6.75	5369.7	4764.2	242.45	5673749.49	1113585.55
				0	0		16.0	189.9	24184.95	30.49		33.5	31.87		166872.3	11096.9	7687.1	6.77	5472.2	5236.8	284.11	6032070.62	1176085.55
				0	0		20.0	197.7	25184.95	31.77		37.5	30.99		176746.1	11617.8	7687.2	6.79	5563.7	5702.9	325.78	6352725.09	1238585.55
ISWB 600	133.7	ISMC 400	484.6	320 ×	10.0	320 ×	20.0	258.5	32931.31	29.30	29.9	36.6	34.56	20.00	250735.1	27977.3	20164.4	9.22	7546.7	7243.6	180.51	8585812.01	2278243.79
					12.0		25.0	276.1	35171.31	30.52	31.5	41.6	34.04		275280.7	29888.8	20710.7	9.22	9018.8	8088.1	225.36	9425438.35	2457443.79
					16.0		32.0	303.7	38691.31	31.36	34.7	48.6	33.80		314274.1	30892.5	21803.0	9.22	9863.7	9298.5	272.22	10659228.48	2739043.79
					20.0		40.0	333.9	42531.31	33.32	37.9	56.6	33.54		357691.1	36169.5	22895.4	9.22	10734.4	10665.2	333.36	11982077.55	3046243.79
		ISMC 350	413	250 ×	10.0	320 ×	20.0	245.7	31303.49	30.68	30.5	36.6	33.13	17.50	237298.3	21473.9	13661.2	8.28	7735.2	7162.1	252.68	8261052.24	1959664.09
					12.0		25.0	262.2	33403.49	32.00	31.9	41.6		32.51	259809.6	23099.7	13921.7	8.32	8120.1	7990.7	303.78	8988241.11	2118914.09
					16.0		32.0	287.7	36643.49	33.49	34.7	48.6		32.12	294693.4	25532.0	14442.7	8.35	8799.5	9174.7	363.14	10045455.04	2360614.09
					20.0		40.0	315.8	40203.49	35.09	37.6	56.6		31.72	333342.5	28237.3	14963.7	8.38	9499.9	10508.6	436.78	11129853.34	2627914.09
		ISMC 400	484.6	0	0	320 ×	10.0	208.3	26531.31	27.79	21.9	26.6	34.07	20.00	184878.1	22516.0	17433.7	9.21	6652.0	5426.9	170.51	6538878.75	1766243.79

Contd...

Composed of								Weight per Metre	Sectional Area a	Centre of Gravity C_{xx}	Mean Thickness of Flanges		Extreme Fibre Distances		Gross Moments of Inertia			Radius of Gyration	Elastic Modulus of Sections		Plastic N-A	Plastic Modulus of Sections	
Joist		Top Flange				Bottom Flange					Top	Bottom	e_{xx}	e_{zz}	I_{xx}	I_{yy}			Z_{zz}				
Design-ation	W	Channel		Plate		Plate										Whole section	Top Flange Only	r_{yy}	Z_1	Z_t	Yp_{zz}	Zp_{xx}	Zp_{yy}
		Design-ation	w	width x	Thick-ness	width x	Thick-ness																
	kg		N	mm	mm	mm	mm	kg	mm²	cm	mm	mm	cm	cm	cm⁴	cm⁴	cm⁴	cm	cm³	cm³	mm	mm³	mm³
				0	0		12.0	213.3	27171.31	28.60		28.6	33.46		192173.4	23062.1	17433.8	9.21	6719.9	5743.0	199.08	6817156.64	1817443.79
				0	0	320	16.0	223.3	28451.31	30.11	21.9	32.6	32.35	20.00	206025.4	24154.4	17434.0	9.21	6841.9	6369.1	256.22	7322695.26	1919843.79
		ISMC 350	413	0	0	320 ×	10.0	201.0	25603.49	28.68	23.3	26.6	33.13	17.50	178422.6	17441.2	12359.0	8.25	6222.0	5384.9	211.43	6380162.63	1547414.09
				0	0		12.0	206.0	26243.49	29.49		28.6	32.52		185319.0	17987.3	12359.1	8.28	6284.8	5698.0	240.00	6631931.36	1598614.09
				0	0		16.0	216.1	27523.49	31.01		32.6	31.40		198388.2	19079.6	12359.3	8.33	6397.9	6317.8	297.14	7084451.68	1701014.09
ISWB 600	145.1	ISMC 400	484.6	320 ×	10.0	320 ×	20.0	269.9	34407.37	29.40	31.4	38.4	34.46	20.00	259894.4	28573.1	20462.3	9.12	8839.0	7542.6	187.53	8959770.04	2355007.98
					12.0		25.0	287.5	36647.37	30.58	33.0	43.4	33.98		284747.6	30484.6	21008.5	9.12	9310.2	8381.0	230.20	9793701.47	2534207.98
					16.0		32.0	315.1	40167.37	31.88	36.2	50.4	33.78		323707.2	33488.3	22100.9	9.13	10152.9	9583.7	274.88	11023888.71	2815807.98
					20.0		40.0	345.2	44007.37	33.31	39.4	58.4	33.55		367122.2	36765.1	23193.3	9.14	11022.5	10941.4	333.12	12345188.17	3123007.98
		ISMC 350	413	250 ×	10.0	320 ×	20.0	257.1	32779.55	30.73	32.1	38.4	33.08	17.50	246744. 1	22069.7	13959.1	8.21	8030.0	7458.5	256.00	8626587.56	2036428.28
					12.0		25.0	273.6	34879.55	32.00	33.5	43.4	32.51		269237.7	23695.5	14219.6	8.25	8414.6	8280.8	304.61	9351490.44	2195678.28
					16.0		32.0	299.0	38119.55	33.45	36.4	50.4	32.16		304137.8	26127.8	14740.6	8.28	9092.7	9456.6	361.15	10409420.72	2437378.28
					20.0		40.0	327.0	41679.55	35.01	39.2	58.4	31.80		342843.2	28833.1	15261.6	8.32	9792.7	10781.2	431.26	11499677.85	2704678.28
		ISMC 400	484.6	0	0	320 ×	10.0	219.6	28007.37	27.95	23.4	28.4	33.91	20.00	194435.4	23111.8	17731.6	9.09	6956.1	5734.1	177.53	6912836.78	1843007.98
				0	0		12.0	224.7	28647.37	28.71		30.4	33.35		201671.9	23657.9	17731.7	9.09	7023.9	6047.5	204.65	7187085.62	1894207.98
				0	0		16.0	234.7	29927.37	30.15		34.4	32.31		215461.2	24750.2	17731.9	9.10	7146.7	6668.2	258.88	7687355.5	1996607.98
		ISMC 350	413	0	0	320 ×	10.0	212.4	27079.55	28.79	25.0	28.4	33.02	17.50	187913.1	18037.0	12656.9	8.17	6526.9	5691.0	216.34	6748582.77	1624178.28
				0	0		12.0	217.4	27719.55	29.50		30.4	32.45		194771.1	18583.1	12657.0	8.19	6589.9	6001.4	243.46	6997670.37	1675378.28
				0	0		16.0	227.4	28999.55	31.00		34.4	31.41		207816.8	19675.4	12657.2	8.24	6704.1	6615.9	297.70	7447617.79	1777778.28
ISMB 550	103.7	ISMC 350	413	250	10.0	320 ×	20.0	215.7	27476.58	28.15	25.7	31.5	30.66	17.50	175306.3	18605.2	12226.8	8.23	6226.9	5718.3	227.68	6629729.57	1668491.71
					12.0		25.0	232.2	29576.58	29.52	27.1	36.5	29.98		194503.9	20231.0	12487.4	8.27	6587.8	6486.7	278.78	7304418.44	1827741.71
					16.0		32.0	257.6	32816.58	31.05	30.0	43.5	29.56		224235.2	22663.3	13008.4	8.31	7222.0	9585.5	338.14	8280632.37	2069441.71
					20.0		40.0	285.6	36376.58	32.66	32.9	51.5	28.15		257183.8	25368.6	13529.3	8.35	7874.2	8823.3	411.78	9276030.67	2336741.71
		ISMC 350	413	0	0	320 ×	10.0	170.9	21776.58	26.05	18.6	21.5	30.76	17.50	125214.8	14572.5	10924.6	8.13	4806.7	4070.7	186.43	4891339.96	1256241.71
				0	0		12.0	176.0	22416.58	26.93		23.5	30.08		131135.8	15118.6	10924.7	8.21	4869.3	4359.7	215.00	5127108.69	1307441.71
				0	0		16.0	186.0	23696.58	28.57		27.5	28.84		142237.4	16210.9	10924.9	8.27	4979.2	4931.1	272.14	5547629.01	1409841.71
		ISMC 300	351.2	0	0	250	10.0	159.2	20274.79	25.68	19.8	24.7	30.98	15.00	114276.0	9498.5	7279.2	6.84	4415.2	3700.9	190.47	4498893.82	981068.74
				0	0		12.0	163.1	20774.79	26.63		26.7	30.33		118958.7	9758.7	7279.3	6.85	4467.4	3921.9	212.79	4682378.36	1012318.74
				0	0		16.0	170.9	21774.79	28.03		30.7	29.33		127852.8	10279.7	7279.5	6.87	4561.3	4359.1	257.43	5018865.29	1074818.74

Contd...

Joist		Top Flange				Bottom Flange		Weight per Metre	Sectional Area a	Centre of Gravity C_{xx}	Mean Thickness of Flanges		Extreme Fibre Distances		Gross Moments of Inertia			Radius of Gyration	Elastic Modulus of Sections		Plastic N-A	Plastic Modulus of Sections	
Design-ation	W	Channel		Plate		Plate					Top	Bottom	e_{xx}	e_{yy}	I_{xx}	I_{yy}		r_{yy}	Z_{zz}		Yp_{zz}	Zp_{xx}	Zp_{yy}
		Design-ation	w	width x	Thick-ness	width x	Thick-ness									Whole section	Top Flange Only		Z_t	Z_t			
	kg		N	mm	mm	mm	mm	kg	mm²	cm	mm	mm	cm	cm	cm⁴	cm⁴	cm⁴	cm	cm³	cm³	mm	mm³	mm³
		ISMC 250	298.2	0	0	250 ×	10.0	153.7	19578.01	26.67	21.8	24.7	30.04	12.50	110209.4	6952.7	4733.5	5.96	4132.7	3668.5	221.08	4372172.33	841021.98
				0	0		12.0	157.6	20078.01	27.42		26.7	29.49		114639.0	17213.1	4733.6	5.99	4181.1	3887.2	243.40	4540103.83	872271.98
				0	0		16.0	165.5	21078.01	28.33		30.7	28.48		123036.8	7733.9	4733.8	6.06	4268.1	4319.7	288.04	4845484.68	934771.98
ISWB 550	112.5	ISMC 400	484.6	320 ×	10.0	320 ×	20.0	237.3	30227.26	26.82	27.6	33.8	32.04	20.00	196231.3	27015.4	19683.5	9.45	7316.2	6124.8	146.30	7253849.24	2159189.26
					12.0		25.0	254.9	32467.26	28.08	29.2	38.8	31.50		217525.8	89926.9	20229.8	9.44	7752.1	6905.6	194.01	8044942.23	2338389.26
					16.0		32.0	282.5	35987.26	29.40	32.4	45.8	31.26		250843.9	31930.6	21322.1	9.42	8532.7	8023.9	243.73	9195456.14	2619989.26
					20.0		40.0	312.5	39827.26	30.65	35.6	53.8	31.01		288010.6	35207.4	22414.5	9.40	9336.9	9286.6	308.68	10424336.93	2927189.26
		ISMC 350	413	250 ×	10.0	320 ×	20.0	224.5	28599.44	28.20	27.8	33.8	30.61	17.50	185333.2	20512.0	13180.3	8.47	6572.4	6054.4	223.31	6980828.07	1840609.56
					12.0		25.0	241.0	30699.44	29.52	29.2	38.8	29.99		204516.4	22137.8	13440.8	8.49	6927.0	6820.5	277.69	7658514.13	1999859.56
					16.0		32.0	266.4	33939.44	31.01	32.1	45.8	29.60		234261.8	24570.1	13961.7	8.51	7554.1	7914.6	340.74	8633788.86	2241559.56
					20.0		40.0	294.4	37499.44	32.59	35.0	53.8	29.22		267256.6	27275.4	14482.7	8.53	8200.2	9146.8	419.03	9621505.58	2508859.56
		ISMC 400	484.6	0	0	320 ×	10.0	187.0	23827.26	25.27	19.6	23.8	31.59	20.00	140570.2	21554.1	16952.8	9.51	5562.9	4449.7	136.30	5366915.99	1647189.26
				0	0		12.0	192.1	24467.26	26.10		25.8	30.96		146829.6	22100.2	16952.9	9.50	5626.0	4742.3	166.77	5634476.39	1698389.26
				0	0		16.0	202.1	25747.26	27.65		29.8	29.81		158641.3	23192.5	16953.0	9.49	5738.0	5321.3	227.73	6114922.93	1800789.26
		ISMC 350	413	0	0	320 ×	10.0	179.8	22899.44	26.16	20.7	23.8	30.89	17.50	135281.9	16479.3	11878.1	8.48	5171.2	4413.9	179.98	5238656.16	1428359.56
				0	0		12.0	184.8	23539.44	27.00		25.8	30.01		141168.7	17025.4	11878.2	8.50	5229.1	4703.5	210.46	5477940.14	1479559.56
				0	0		16.0	194.8	24819.44	28.55		29.8	28.86		152250.6	18117.7	11878.3	8.54	5331.9	5276.4	271.41	5901833.81	1581959.56
ISMB 500	86.9	ISMC 350	413	250 ×	10.0	320 ×	20.0	198.9	25339.88	25.70	24.5	29.7	28.11	17.50	136749.0	18141.2	11994.9	8.40	5320.9	4864.8	196.26	5631084.35	1600051.60
					12.0		25.0	215.4	27439.88	27.06	25.5	34.7	27.45		152843.0	19767.0	12255.4	8.49	5649.5	5569.0	252.18	6257680.86	1759301.60
					16.0		32.0	240.8	30679.88	28.56	28.4	41.7	27.05		177877.9	22199.0	12776.3	8.51	6228.8	6575.2	316.97	7151513.61	2001001.60
					20.0		40.0	268.3	34239.88	30.13	31.2	49.7	26.68		205641.6	24904.6	13297.3	8.53	6825.0	7707.9	397.44	8046615.47	2268301.60
		ISMC 350	413	0	0	320 ×	10.0	154.2	19639.88	23.58	17.0	19.7	28.23	17.50	94789.4	14108.5	10692.7	8.48	4020.0	3357.7	151.95	4029632.54	1187801.60
				0	0		12.0	159.2	20279.88	24.47		21.7	27.54		99764.0	14654.6	10692.8	8.50	4076.4	3623.0	183.32	4254570.75	1239001.60
				0	0		16.0	169.2	21559.88	26.12		25.7	26.29		109026.6	15746.9	10692.9	8.55	4174.0	4147.1	246.07	4648051.89	1341401.60
		ISMC 300	351.2	0	0	250 ×	10.0	142.4	18139.09	23.39	17.9	22.4	28.37	15.00	85756.4	9034.5	7047.3	7.06	3666.9	3022.4	156.44	3675673.51	912628.63
				0	0		12.0	146.3	18639.09	24.15		24.4	27.81		89701.3	9294.9	7047.4	7.06	3714.3	3225.6	180.95	3850627.12	943878.63
				0	0		16.0	154.2	19638.09	25.58		28.4	26.78		97147.3	9815.7	7047.5	7.07	3798.3	3627.2	229.97	4166769.63	1006378.63
		ISMC 250	298.2	0	0	250 ×	10.0	136.9	17441.31	24.17	19.5	22.4	27.54	12.50	82476.9	6488.7	4501.6	6.10	3411.7	2995.3	190.09	3571602.49	772581.87
				0	0		12.0	140.8	17941.31	24.94		24.4	26.97		86189.0	6749.1	4501.6	6.13	3455.2	3196.3	214.00	3729478.25	803831.87
				0	0		16.0	148.7	18941.31	26.38		28.4	25.93		93179.0	7269.9	4501.8	6.20	3532.3	3593.4	263.62	4011465.07	866331.87

Contd...

Composed of: Joist Designation	W	Top Flange: Channel Designation	Top Flange: Channel w	Top Flange: Plate width x	Top Flange: Plate Thickness	Bottom Flange: Plate width x	Bottom Flange: Plate Thickness	Weight per Metre	Sectional Area a	Centre of Gravity C_{xx}	Mean Thickness of Flanges: Top	Mean Thickness of Flanges: Bottom	Extreme Fibre Distances: e_{xx}	Extreme Fibre Distances: e_{yy}	Gross Moments of Inertia: I_{xx}	Gross Moments of Inertia: I_{yy} Whole section	Gross Moments of Inertia: I_{yy} Top Flange Only	Radius of Gyration r_{yy}	Elastic Modulus of Sections Z_{zz}: Z_1	Elastic Modulus of Sections Z_{zz}: Z_1	Plastic N-A Yp_{zz}	Plastic Modulus of Sections: Zp_{xx}	Plastic Modulus of Sections: Zp_{yy}
	kg		N	mm	mm	mm	mm	kg	mm²	cm	mm	mm	cm	cm	cm⁴	cm⁴	cm⁴	cm	cm³	cm³	mm	mm³	mm³
ISWB 500	95.2	ISMC 400	484.6	320 ×	10.0	320 ×	20.0	219.9	28015.24	24.40	25.8	31.5	28.46	20.00	152781.2	26262.6	19307.2	9.68	6262.3	5185.5	112.37	6129806.87	2065841.75
					12.0		25.0	237.5	30255.24	25.63	27.4	36.5	28.93		170717.9	27174.1	19853.4	9.65	6661.7	5900.4	162.86	6872140.24	2245041.75
					15.0		32.0	265.1	33775.24	26.95	30.6	43.5	28.71		198797.6	31177.8	20945.7	9.61	7377.1	6923.9	215.34	7939234.80	2526641.75
					20.0		40.0	295.3	37615.24	28.37	33.8	51.5	28.49		230194.5	34454.6	22038.1	9.57	8114.6	8079.2	283.99	9074085.74	2833841.75
		ISMC 350	413.0	250 ×	10.0	320 ×	20.0	207.1	26387.42	25.74	25.7	31.5	28.07	17.50	143834.0	19759.2	12803.9	8.65	5587.0	5124.9	194.08	5908189.66	1747262.04
					12.0		25.0	223.6	28487.42	27.06	27.2	36.5	27.45		159945.7	21385.0	13064.4	8.66	5911.3	5826.3	251.64	6536282.09	1906512.04
					15.0		32.0	249.1	31727.42	28.52	30.0	43.5	27.09		184963.8	23817.3	13585.3	8.66	6485.6	6827.6	318.27	7429646.08	2148212.04
					20.0		40.0	277.0	35287.42	30.08	32.9	51.5	26.75		212768.9	26522.6	14106.2	8.67	7077.7	7954.5	401.05	8320913.99	2415512.04
		ISMC 400	484.6	0	0	320 ×	10.0	169.7	21615.24	22.61	17.8	21.5	29.05	20.00	106172.6	20801.3	16576.5	9.81	4654.5	3654.9	102.37	4402873.61	1553841.75
				0	0		12.0	174.7	22255.24	23.65		23.5	28.41		111454.1	21347.4	16576.5	9.79	4712.8	3922.9	134.69	4659556.47	1605041.75
				0	0		16.0	184.7	23535.24	25.21		27.5	27.25		121362.3	22439.7	16576.6	9.76	4815.0	4452.9	199.34	5114701.58	1707441.75
		ISMC 350	413	0	0	320 ×	10.0	162.4	20687.42	23.69	21.5	21.5	28.12	17.50	101911.5	15726.5	11501.7	8.72	4301.4	3654.5	148.73	4304850.07	1335012.04
				0	0		12.0	167.4	21327.42	24.54	18.6	23.5	27.47		106854.4	16272.6	11501.8	8.73	4354.4	3889.8	181.05	4531542.77	1386212.04
				0	0		16.0	177.5	22607.42	26.11	27.5	27.5	26.30		116100.2	17364.9	11501.9	8.76	4447.3	4413.8	245.70	4926707.57	1488612.04
ISMB 450	72.4	ISMC 300	351.2	0	0	250 ×	10.0	127.9	16290.34	20.93	16.3	20.4	25.83	15.00	62983.5	8498.7	6779.4	7.22	3008.6	2438.8	122.83	2968893.99	840146.29
				0	0		12.0	131.8	16790.34	21.71		22.4	25.25		66244.2	8759.1	6779.5	7.22	3051.8	2623.1	149.42	3135130.88	871396.29
				0	0		16.0	139.7	17790.34	23.14		26.4	24.22		72359.1	9279.9	6779.6	7.22	3127.4	2987.2	202.62	3430711.05	933896.29
		ISMC 250	298.2	0	0	250 ×	10.0	122.4	15593.56	21.71	17.5	20.4	25.00	12.50	60394.7	5952.9	4233.7	6.18	2781.6	2416.0	159.39	2887228.58	700099.53
				0	0		12.0	126.3	16093.56	22.49		22.4	24.42		63446.4	6213.3	4233.8	6.21	2820.9	2598.3	185.99	3034934.19	731349.53
				0	0		16.0	134.2	17093.56	23.93		26.4	23.38		69152.5	6734.1	4233.8	6.28	2889.5	2958.0	239.18	3293451.80	793849.53
		ISMC 225	254.1	0	0	250 ×	10.0	114.0	15028.06	21.57	18.0	23.0	25.07	11.25	55138.5	4195.3	3111.5	5.37	2556.1	2199.5	188.77	2799283.46	621307.02
				0	0		12.0	117.2	15528.06	22.25		25.0	24.59		57604.5	4328.6	3111.5	5.38	2589.5	2342.2	216.37	2931949.18	652557.02
				0	0		16.0	123.5	16528.06	23.51		29.0	23.73		62272.6	4595.3	3111.6	5.41	2649.1	2623.8	268.56	3160387.00	715057.02
ISWB 450	79.4	ISMC 350	413	250 ×	10.0	320 ×	20.0	191.4	24380.56	23.31	24.0	29.6	25.50	17.50	109500.7	18478.1	12163.4	8.71	4698.2	4293.6	163.45	4954650.64	1624615.84
					12.0		25.0	207.9	26480.56	24.60	25.5	34.6	24.91		122801.8	20103.9	12423.9	8.71	4992.8	4929.0	225.24	5534112.96	1783865.84
					16.0		32.0	233.3	29720.56	26.02	28.3	41.6	24.59		143501.0	22536.2	12944.8	8.71	5514.9	5835.9	296.63	6345264.27	2025565.84
					20.0		40.0	261.3	33280.56	27.51	31.2	49.6	24.30		166568.3	25241.5	13465.7	8.71	6053.8	6856.0	385.41	7137613.93	2292865.84
		ISMC 350	413.0	0	0	320 ×	10.0	146.6	18680.56	21.26	16.9	19.6	25.55	17.50	75002.0	14445.4	10861.2	8.79	3528.7	2934.9	115.41	3488927.44	1212365.84
				0	0		12.0	151.7	19320.56	22.10		21.6	24.91		79074.8	14991.5	10861.3	8.81	3577.3	3175.0	150.10	3704158.93	1263565.84
				0	0		16.0	161.7	20600.56	23.66		25.6	23.75		86841.2	16083.8	10861.4	8.84	3661.2	3648.8	219.76	4071679.32	1365965.84
		ISMC 300	351.2	0	0	250 ×	10.0	134.9	17178.76	21.05	17.9	22.3	25.71	15.00	67695.9	9371.4	7215.8	7.39	3215.3	2633.5	120.44	3173660.34	937192.87
				0	0		12.0	138.8	17678.76	21.78		24.3	25.18		70931.4	9631.8	7215.9	7.38	3256.1	2817.5	147.62	3340945.86	968442.87
				0	0		16.0	146.6	18678.76	23.14		28.3	24.22		77026.0	10152.6	7215.9	7.37	3328.3	3180.6	201.96	3637756.03	1030942.87

Tensile Load Capacity of Single Equal Angle Sections

Table VIII Tensile Load Capacity of Equal Angle Sections for Yield Stress and Ultimate Stress With Welded end Connections

Section	th mm	Area (mm²)	Ultimate Strength (kN)	Yield Strength (kN)
ISA 2020 @ 0.9 kg/m	3	111.751	32.989	25.398
ISA 2020 @ 1.1 kg/m	4	144.751	42.731	32.898
ISA 2525 @ 1.1 kg/m	3	141.483	41.766	32.155
ISA 2525 @ 1.4 kg/m	4	184.483	54.459	41.928
ISA 2525 @ 1.8 kg/m	5	225.483	66.563	51.246
ISA 3030 @ 1.4 kg/m	3	172.502	50.923	39.205
ISA 3030 @ 1.8 kg/m	4	225.502	66.568	51.251
ISA 3030 @ 2.2 kg/m	5	276.502	81.623	62.841
ISA 3535 @ 1.6 kg/m	3	202.502	59.779	46.023
ISA 3535 @ 2.1 kg/m	4	265.502	78.376	60.341
ISA 3535 @ 2.6 kg/m	5	326.502	96.383	74.205
ISA 3535 @ 3.0 kg/m	6	385.502	113.800	87.614
ISA 4040 @ 1.8 kg/m	3	233.629	68.967	53.097
ISA 4040 @ 2.4 kg/m	4	306.629	90.517	69.688
ISA 4040 @3.0 kg/m	5	377.629	111.476	85.825
ISA 4040 @ 3.5kg/m	6	446.629	131.845	101.507
ISA 4545 @ 2.1 kg/m	3	263.629	77.823	59.916
ISA 4545 @ 2.7 kg/m	4	346.629	102.325	78.779
ISA 4545 @ 3.4 kg/m	5	427.629	126.236	97.188
ISA 4545 @ 4.0 kg/m	6	506.629	149.557	115.143
ISA 5050 @ 2.3 kg/m	3	294.863	87.044	67.014
ISA 5050 @3.0 kg/m	4	387.863	114.497	88.151
ISA 5050 @3.8 kg/m	5	478.863	141.360	108.832
ISA 5050 @ 4.5 kg/m	6	567.863	167.633	129.060
ISA 5555 @4.1 kg/m	5	527.200	155.629	119.818
ISA 5555 @ 4.9 kg/m	6	626.200	184.854	142.318
ISA 5555 @ 6.4 kg/m	8	818.200	241.533	185.954
ISA 5555 @7.9 kg/m	10	1002.200	295.849	227.773
ISA 6060 @ 4.5 kg/m	5	575.376	169.851	130.767
ISA 6060 @ 5.4 kg/m	6	684.376	202.028	155.540
ISA 6060 @ 7.0 kg/m	8	896.376	264.610	203.722
ISA 6060 @ 8.6 kg/m	10	1100.376	324.831	250.085
ISA 6565 @ 4.9 kg/m	5	625.376	184.611	142.131
ISA 6565 @ 5.8 kg/m	6	744.376	219.740	169.176
ISA 6565 @ 7.7 kg/m	8	976.376	288.226	221.904
ISA 6565 @ 9.4 kg/m	10	1200.376	354.351	272.813
ISA 7070 @ 5.3 kg/m	5	676.824	199.798	153.824
ISA 7070 @ 6.3 kg/m	6	805.824	237.879	183.142
ISA 7070 @ 8.3 kg/m	8	1057.824	312.270	240.415
ISA 7070 @ 10.2 kg/m	10	1301.824	384.298	295.869
ISA 7575 @ 5.7 kg/m	5	726.824	214.558	165.187
ISA 7575 @6.8 kg/m	6	865.824	255.591	196.778

Contd...

SA 7575 @ 8.9 kg/m	8	1137.824	335.886	258.596
SA 7575 @ 11.0 kg/m	10	1401.824	413.818	318.596
SA 8080 @ 7.3 kg/m	6	929.043	274.254	211.146
SA 8080 @ 9.6 kg/m	8	1221.043	360.452	277.510
SA 8080 @ 11.8 kg/m	10	1505.043	444.289	342.055
SA 8080 @ 14.0 kg/m	12	1781.043	525.764	404.783
SA 9090 @ 8.2 kg/m	6	1046.522	308.933	237.846
SA 9090 @ 10.8 kg/m	8	1378.522	406.940	313.300
SA 9090 @ 13.4 kg/m	10	1702.522	502.584	386.937
SA 9090 @ 15.8 kg/m	12	2018.522	595.868	458.755
SA 100100 @ 9.2 kg/m	6	1166.522	344.357	265.119
SA 100100 @ 12.1 kg/m	8	1538.522	454.172	349.664
SA 100100 @ 14.9 kg/m	10	1902.522	561.624	432.391
SA 100100 @ 17.7 kg/m	12	2258.522	666.716	513.300
SA 110110 @ 13.4 kg/m	8	1702.009	502.433	386.820
SA 110110 @ 16.5 kg/m	10	2106.009	621.694	478.638
SA 110110 @ 19.6 kg/m	12	2502.009	738.593	568.638
SA 110110 @ 24.2 kg/m	15	3081.009	909.514	700.229
SA 130130 @ 15.9 kg/m	8	2022.009	596.897	459.547
SA 130130 @ 19.7 kg/m	10	2506.009	739.774	569.547
SA 130130 @ 23.4 kg/m	12	2982.009	880.289	677.729
SA 130130 @ 28.9 kg/m	15	3681.009	1086.634	836.593
SA 150150 @ 22.8 kg/m	10	2903.434	857.094	659.871
SA 150150 @ 27.2 kg/m	12	3459.434	1021.225	786.235
SA 150150 @33.6 kg/m	15	4278.434	1262.994	972.371
SA 150150 @ 39.9 kg/m	18	5079.434	1499.449	1154.417
SA 200200 @ 36.6 kg/m	12	4661.365	1376.035	1059.401
SA 200200 @ 45.4 kg/m	15	5780.365	1706.364	1313.719
SA 200200 @ 54.0 kg/m	18	6881.365	2031.379	1563.947
SA 200200 @ 73.6kg/m	25	9380.365	2769.084	2131.901

Tensile Load Capacity of Single Unequal Angle Sections

able IX Tensile Load Capacity of Unequal Angle Sections for Yield Stress and Ultimate Stress with Welded end Connections

Section	th mm	Area (mm^2)	Ultimate Strength (kN)	Yield Strength (kN)
ISA 3020 @ 1.1 kg/m	3	141.483	41.766	32.155
ISA 3020 @ 1.1 kg/m	4	184.483	54.459	41.928
ISA 3020 @ 1.1 kg/m	5	225.483	66.563	51.246
ISA 4025 @ 1.5 kg/m	3	187.502	55.351	42.614
ISA 4025 @ 1.9 kg/m	4	245.502	72.472	55.796
ISA 4025 @ 2.4 kg/m	5	301.502	89.003	68.523
ISA 4025 @ 2.8 kg/m	6	355.502	104.944	80.796
ISA 4530 @ 1.7 kg/m	3	217.502	64.207	49.432
ISA 4530 @ 2.2 kg/m	4	285.502	84.280	64.887
ISA 4530 @ 2.8 kg/m	5	351.502	103.763	79.887
ISA 4530 @ 3.3 kg/m	6	415.502	122.656	94.432
ISA 5030 @ 1.8 kg/m	3	233.629	68.967	53.097
SA 5030 @ 2.4 kg/m	4	306.629	90.517	69.688
ISA 5030 @ 3.0 kg/m	5	377.629	111.476	85.825
SA 5030 @ 3.5 kg/m	6	446.629	131.845	101.507
SA 6040 @ 3.7 kg/m	5	475.858	140.473	108.150
SA 6040 @ 4.4 kg/m	6	564.858	166.746	128.377
SA 6040 @ 5.8 kg/m	8	736.858	217.521	167.468
SA 6545 @ 4.1 kg/m	5	525.858	155.233	119.513
SA 6545 @ 4.9 kg/m	6	624.858	184.458	142.013
SA 6545 @ 6.4 kg/m	8	816.858	241.137	185.650
SA 7045 @ 4.3 kg/m	5	552.200	163.009	125.500
SA 7045 @ 5.2 kg/m	6	656.200	193.710	149.136
SA 7045 @ 6.7 kg/m	8	858.200	253.341	195.045
SA 7045 @ 8.3 kg/m	10	1052.200	310.609	239.136
SA 7550 @ 4.7 kg/m	5	602.200	177.769	136.864
SA 7550 @ 5.6 kg/m	6	716.200	211.422	162.773
SA 7550 @ 7.4 kg/m	8	938.200	276.957	213.227
SA 7550 @ 9.0 kg/m	10	1152.200	340.129	261.864
SA 8050 @ 4.9 kg/m	5	626.824	185.038	142.460
SA 8050 @ 5.9 kg/m	6	745.824	220.167	169.505
SA 8050 @ 7.7 kg/m	8	977.824	288.654	222.233
SA 8050 @ 9.4 kg/m	10	1201.824	354.778	273.142
SA 9060 @ 6.8 kg/m	6	865.341	255.449	196.668
SA 9060 @ 8.9 kg/m	8	1137.341	335.743	258.487
SA 9060 @ 11.0 kg/m	10	1401.341	413.676	318.487
SA 9060 @ 13.0 kg/m	12	1657.341	489.247	376.668
SA 10065 @ 7.5 kg/m	6	954.751	281.843	216.989
SA 10065 @ 9.9 kg/m	8	1256.751	370.993	285.625
SA 10065 @ 12.2 kg/m	10	1550.751	457.782	352.443
SA 10075 @ 8.0 kg/m	6	1014.054	299.349	230.467
SA 10075 @ 10.5 kg/m	8	1336.054	394.403	303.649

Contd...

ISA 10075 @ 13.0 kg/m	10	1650.054	487.096	375.012
ISA 10075 @ 15.4 kg/m	12	1956.054	577.427	444.558
ISA 12575 @ 9.2 kg/m	6	1165.931	344.183	264.984
ISA 12575 @ 12.1 kg/m	8	1537.931	453.997	349.530
ISA 12575 @ 14.9 kg/m	10	1901.931	561.450	432.257
ISA 12595 @ 10.1 kg/m	6	1285.931	379.607	292.257
ISA 12595 @ 13.3 kg/m	8	1697.931	501.229	385.894
ISA 12595 @ 16.5 kg/m	10	2101.931	620.490	477.712
ISA 12595 @ 19.6 kg/m	12	2497.931	737.389	567.712
ISA 15075 @ 13.7 kg/m	8	1742.009	514.241	395.911
ISA 15075 @ 16.9 kg/m	10	2156.009	636.454	490.002
ISA 15075 @ 20.1 kg/m	12	2562.009	756.305	582.275
ISA 150115 @ 16.2 kg/m	8	2057.824	607.470	467.687
ISA 150115 @ 20.0 kg/m	10	2551.824	753.298	579.960
ISA 150115 @ 23.8 kg/m	12	3037.824	896.766	690.415
ISA 150115 @ 29.5 kg/m	15	3751.824	1107.538	852.687
ISA 200100 @ 22.8 kg/m	10	2903.434	857.094	659.871
ISA 200100 @ 27.2 kg/m	12	3459.434	1021.225	786.235
ISA 200100 @ 33.6 kg/m	15	4278.434	1262.994	972.371
ISA 200150 @ 26.7 kg/m	10	3400.376	1003.791	772.813
ISA 200150 @ 31.8 kg/m	12	4056.376	1197.442	921.904
ISA 200150 @ 39.4 kg/m	15	5025.376	1483.491	1142.131
ISA 200150 @ 46.9 kg/m	18	5976.376	1764.226	1358.267

Factored Axial Compression Load Capacity of Single Equal Angle Section

Table X (01) Axial Compression Load Capacity of Single Equal Angle Section

ISA 2020(3)@ 0.9 kg/m

Section	Area mm^2	$r_{min} = r_{vv}$ mm	Buckling Class y-y
ISA 2020(3)@ 0.9 kg/m	111.751	3.70	c
Eff Length m	Sl Ratio	D.C.S. f_{cd} N/mm^2	F.A.L. kN
0.100	27.027	214.865	24.011
0.200	54.054	176.919	19.771
0.300	81.081	134.378	15.017
0.400	108.108	96.946	10.834
0.500	135.135	70.141	7.838
0.600	162.162	52.176	5.831
0.700	189.189	40.016	4.472
0.800	216.216	31.622	3.534
0.900	243.243	25.584	2.859
0.925	250.000	24.300	2.716

TABLE X (02) Axial Compression Load Capacity of Single Equal Angle Section

ISA 2020(4)@ 1.1 kg/m

Section	Area mm^2	$r_{min} = r_{vv}$ mm	Buckling Class y-y
ISA 2020(4)@ 1.1 kg/m	111.751	3.70	c
Eff Length m	Sl Ratio	D.C.S. f_{cd} N/mm^2	F.A.L. kN
0.100	27.027	214.865	31.102
0.200	54.054	176.919	25.609
0.300	81.081	134.378	19.451
0.400	108.108	96.946	14.033
0.500	135.135	70.141	10.153
0.600	162.162	52.176	7.552
0.700	189.189	40.016	5.792
0.800	216.216	31.622	4.577
0.900	243.243	25.584	3.703
0.925	250.000	24.300	3.517

Table X (03) Axial Compression Load Capacity of Single Equal Angle Section

ISA 2525(3)@ 1.1 kg/m

Section	Area mm^2	$r_{min} = r_{vv}$ mm	Buckling Class y-y
ISA 2525(3)@ 1.1 kg/m	144.751	4.70	c
Eff Length m	Sl Ratio	D.C.S. f_{cd} N/mm^2	F.A.L. kN
1.000	212.766	32.553	4.606
1.100	234.043	27.451	3.884
1.175	250.000	24.300	3.438

able X (04) Axial Compression Load Capacity of Single Equal Angle Section

ISA 2525(4)@ 1.4 kg/m

Section	Area mm^2	$r_{min} = r_{vv}$ mm	Buckling Class y-y
ISA 2525(4)@ 1.4 kg/m	184.482	4.70	c
Eff Length m	Sl Ratio	D.C.S. f_{cd} N/mm^2	F.A.L. kN
1.000	212.766	32.553	6.006
1.100	234.043	27.451	5.064
1.175	250.000	24.300	4.483

able X (05) Axial Compression Load Capacity of Single Equal Angle Section

ISA 2525(5)@ 1.8 kg/m

Section	Area mm^2	$r_{min} = r_{vv}$ mm	Buckling Class y-y
ISA 2525(5)@ 1.8 kg/m	225.482	4.70	c
Eff Length m	Sl Ratio	D.C.S. f_{cd} N/mm^2	F.A.L. kN
1.000	212.766	32.553	7.340
1.100	234.043	27.451	6.190
1.175	250.000	24.300	5.479

able X (06) Axial Compression Load Capacity of Single Equal Angle Section

ISA 3030(3)@ 1.4 kg/m

Section	Area mm^2	$r_{min} = r_{vv}$ mm	Buckling Class y-y
ISA 3030(3)@ 1.4 kg/m	172.502	5.70	c
Eff Length m	Sl Ratio	D.C.S. f_{cd} N/mm^2	F.A.L. kN
1.000	175.439	45.653	7.875
1.100	192.982	38.686	6.673
1.200	210.526	33.158	5.720
1.300	228.070	28.744	4.958
1.400	245.614	25.133	4.336
1.425	250.000	24.300	4.192

able X (07) Axial Compression Load Capacity of Single Equal Angle Section

ISA 3030(4)@ 1.8 kg/m

Section	Area mm^2	$r_{min} = r_{vv}$ mm	Buckling Class y-y
ISA 3030(4)@ 1.8 kg/m	225.502	5.70	c
Eff Length m	Sl Ratio	D.C.S. f_{cd} N/mm^2	F.A.L. kN
1.000	175.439	45.653	10.295
1.100	192.982	38.686	8.724
1.200	210.526	33.158	7.477
1.300	228.070	28.744	6.482
1.400	245.614	25.133	5.668
1.425	250.000	24.300	5.480

Table X (08) Axial Compression Load Capacity of Single Equal Angle Section

ISA 3030(5)@ 2.2 kg/m

Section	*Area mm²*	$r_{min} = r_{vv}$ *mm*	*Buckling Class y-y*
ISA 3030(5)@ 2.2 kg/m	*276.502*	*5.70*	*c*
Eff Length m	*Sl Ratio*	*D.C.S.* f_{cd} *N/mm²*	*F.A.L. kN*
1.000	175.439	45.653	12.623
1.100	192.982	38.686	10.697
1.200	210.526	33.158	9.168
1.300	228.070	28.744	7.948
1.400	245.614	25.133	6.949
1.425	250.000	24.300	6.719

Table X (09) Axial Compression Load Capacity of Single Equal Angle Section

ISA 3535(3)@ 1.6 kg/m

Section	*Area mm²*	$r_{min} = r_{vv}$ *mm*	*Buckling Class y-y*
ISA 3535(3)@ 1.6 kg/m	*202.502*	*6.70*	*c*
Eff Length m	*Sl Ratio*	*D.C.S.* f_{cd} *N/mm²*	*F.A.L. kN*
1.000	149.254	59.722	12.094
1.100	164.179	51.127	10.353
1.200	179.104	44.003	8.911
1.300	194.030	38.330	7.762
1.400	208.955	33.613	6.807
1.500	223.881	29.707	6.016
1.600	238.806	26.451	5.356
1.675	250.000	24.300	4.921

Table X (10) Axial Compression Load Capacity of Single Equal Angle Section

ISA 3535(4)@ 2.1 kg/m

Section	*Area mm²*	$r_{min} = r_{vv}$ *mm*	*Buckling Class y-y*
ISA 3535(4)@ 2.1 kg/m	*265.502*	*6.70*	*c*
Eff Length m	*Sl Ratio*	*D.C.S.* f_{cd} *N/mm²*	*F.A.L. kN*
1.000	149.254	59.722	15.856
1.100	164.179	51.127	13.574
1.200	179.104	44.003	11.683
1.300	194.030	38.330	10.177
1.400	208.955	33.613	8.924
1.500	223.881	29.707	7.887
1.600	238.806	26.451	7.023
1.675	250.000	24.300	6.452

able X (11) Axial Compression Load Capacity of Single Equal Angle Section

ISA 3535(5)@ 2.6 kg/m

Section	Area mm^2	$r_{min} = r_{vv}$ mm	Buckling Class y-y
ISA 3535(5)@ 2.6 kg/m	326.502	6.70	c
Eff Length m	Sl Ratio	D.C.S. f_{cd} N/mm²	F.A.L. kN
1.000	149.254	59.722	19.499
1.100	164.179	51.127	16.693
1.200	179.104	44.003	14.367
1.300	194.030	38.330	12.515
1.400	208.955	33.613	10.975
1.500	223.881	29.707	9.700
1.600	238.806	26.451	8.636
1.675	250.000	24.300	7.934

able X (12) Axial Compression Load Capacity of Single Equal Angle Section

ISA 3535(6)@ 3.0 kg/m

Section	Area mm^2	$r_{min} = r_{vv}$ mm	Buckling Class y-y
ISA 3535(6)@ 3.0 kg/m	385.502	6.70	c
Eff Length m	Sl Ratio	D.C.S. f_{cd} N/mm²	F.A.L. kN
1.000	149.254	59.722	23.023
1.100	164.179	51.127	19.710
1.200	179.104	44.003	16.963
1.300	194.030	38.330	14.776
1.400	208.955	33.613	12.958
1.500	223.881	29.707	11.452
1.600	238.806	26.451	10.197
1.675	250.000	24.300	9.368

able X (13) Axial Compression Load Capacity of Single Equal Angle Section

ISA 4040(3)@ 1.8 kg/m

Section	Area mm^2	$r_{min} = r_{vv}$ mm	Buckling Class y-y
ISA 4040(3)@ 1.8 kg/m	233.628	7.70	c
Eff Length m	Sl Ratio	D.C.S. f_{cd} N/mm²	F.A.L. kN
1.000	129.870	74.422	17.387
1.100	142.857	64.200	14.999
1.200	155.844	55.752	13.025
1.300	168.831	48.708	11.380
1.400	181.818	42.891	10.021
1.500	194.805	38.066	8.893
1.600	207.792	33.962	7.935
1.700	220.779	30.421	7.107
1.800	233.766	27.509	6.427
1.900	246.753	24.917	5.821
1.925	250.000	24.300	5.677

Table X (14) Axial Compression Load Capacity of Single Equal Angle Section

ISA 4040(4)@ 2.4 kg/m

Section	Area mm²	$r_{min} = r_{vv}$ mm	Buckling Class y-y
ISA 4040(4)@ 2.4 kg/m	306.628	7.70	c
Eff Length m	Sl Ratio	D.C.S. f_{cd} N/mm²	F.A.L. kN
1.000	129.870	74.422	22.820
1.100	142.857	64.200	19.686
1.200	155.844	55.752	17.095
1.300	168.831	48.708	14.935
1.400	181.818	42.891	13.152
1.500	194.805	38.066	11.672
1.600	207.792	33.962	10.414
1.700	220.779	30.421	9.328
1.800	233.766	27.509	8.435
1.900	246.753	24.917	7.640
1.925	250.000	24.300	7.451

Table X (15) Axial Compression Load Capacity of Single Equal Angle Section

ISA 4040(5)@3.0 kg/m

Section	Area mm²	$r_{min} = r_{vv}$ mm	Buckling Class y-y
ISA 4040(5)@3.0 kg/m	377.628	7.70	c
Eff Length m	Sl Ratio	D.C.S. f_{cd} N/mm²	F.A.L. kN
1.000	129.870	74.422	28.104
1.100	142.857	64.200	24.244
1.200	155.844	55.752	21.054
1.300	168.831	48.708	18.393
1.400	181.818	42.891	16.197
1.500	194.805	38.066	14.375
1.600	207.792	33.962	12.825
1.700	220.779	30.421	11.488
1.800	233.766	27.509	10.388
1.900	246.753	24.917	9.409
1.925	250.000	24.300	9.176

Table X (16) Axial Compression Load Capacity of Single Equal Angle Section

ISA 4040(6)@ 3.5kg/m

Section	Area mm²	$r_{min} = r_{vv}$ mm	Buckling Class y-y
ISA 4040(6)@ 3.5kg/m	446.628	7.70	c
Eff Length m	Sl Ratio	D.C.S. f_{cd} N/mm²	F.A.L. kN
1.000	129.870	74.422	33.239
1.100	142.857	64.200	28.674
1.200	155.844	55.752	24.900
1.300	168.831	48.708	21.754
1.400	181.818	42.891	19.156

Contd...

1.500	194.805	38.066	17.001
1.600	207.792	33.962	15.169
1.700	220.779	30.421	13.587
1.800	233.766	27.509	12.286
1.900	246.753	24.917	11.129
1.925	250.000	24.300	10.853

able X (17) Axial Compression Load Capacity of Single Equal Angle Section

ISA 4545(3)@ 2.1 kg/m

Section	*Area mm²*	$r_{min} = r_{vv}$ *mm*	*Buckling Class y-y*
ISA 4545(3)@ 2.1 kg/m	*263.628*	*8.70*	*c*
Eff Length m	*Sl Ratio*	*D.C.S.* f_{cd} *N/mm²*	*F.A.L. kN*
1.000	114.943	89.213	23.519
1.100	126.437	77.649	20.471
1.200	137.931	67.876	17.894
1.300	149.425	59.602	15.713
1.400	160.920	52.822	13.925
1.500	172.414	47.014	12.394
1.600	183.908	42.076	11.092
1.700	195.402	37.863	9.982
1.800	206.897	34.231	9.024
1.900	218.391	31.034	8.182
2.000	229.885	28.326	7.468
2.100	241.379	25.938	6.838
2.175	250.000	24.300	6.406

able X (18) Axial Compression Load Capacity of Single Equal Angle Section

ISA 4545(4)@ 2.7 kg/m

Section	*Area mm²*	$r_{min} = r_{vv}$ *mm*	*Buckling Class y-y*
ISA 4545(4)@ 2.7 kg/m	*346.628*	*8.70*	*c*
Eff Length m	*Sl Ratio*	*D.C.S.* f_{cd} *N/mm²*	*F.A.L. kN*
1.000	114.943	89.213	30.924
1.100	126.437	77.649	26.916
1.200	137.931	67.876	23.528
1.300	149.425	59.602	20.660
1.400	160.920	52.822	18.310
1.500	172.414	47.014	16.296
1.600	183.908	42.076	14.585
1.700	195.402	37.863	13.124
1.800	206.897	34.231	11.865
1.900	218.391	31.034	10.757
2.000	229.885	28.326	9.819
2.100	241.379	25.938	8.991
2.175	250.000	24.300	8.423

Table X (19) Axial Compression Load Capacity of Single Equal Angle Section

ISA 4545(5)@ 3.4 kg/m

Section	*Area mm²*	$r_{min} = r_{vv}$ *mm*	*Buckling Class y-y*
ISA 4545(5)@ 3.4 kg/m	*427.628*	*8.70*	*c*
Eff Length m	*Sl Ratio*	*D.C.S.* f_{cd} *N/mm²*	*F.A.L. kN*
1.000	114.943	89.213	38.150
1.100	126.437	77.649	33.205
1.200	137.931	67.876	29.026
1.300	149.425	59.602	25.488
1.400	160.920	52.822	22.588
1.500	172.414	47.014	20.104
1.600	183.908	42.076	17.993
1.700	195.402	37.863	16.191
1.800	206.897	34.231	14.638
1.900	218.391	31.034	13.271
2.000	229.885	28.326	12.113
2.100	241.379	25.938	11.092
2.175	250.000	24.300	10.391

Table X (20) Axial Compression Load Capacity of Single Equal Angle Section

ISA 4545(6)@ 4.0 kg/m

Section	*Area mm²*	$r_{min} = r_{vv}$ *mm*	*Buckling Class y-y*
ISA 4545(6)@ 4.0 kg/m	*506.628*	*8.70*	*c*
Eff Length m	*Sl Ratio*	*D.C.S.* f_{cd} *N/mm²*	*F.A.L. kN*
1.000	114.943	89.213	45.198
1.100	126.437	77.649	39.339
1.200	137.931	67.876	34.388
1.300	149.425	59.602	30.196
1.400	160.920	52.822	26.761
1.500	172.414	47.014	23.819
1.600	183.908	42.076	21.317
1.700	195.402	37.863	19.183
1.800	206.897	34.231	17.342
1.900	218.391	31.034	15.723
2.000	229.885	28.326	14.351
2.100	241.379	25.938	13.141
2.175	250.000	24.300	12.311

able X (21) Axial Compression Load Capacity of Single Equal Angle Section

ISA 5050(3)@ 2.3 kg/m

Section	*Area mm²*	$r_{min} = r_{vv}$ *mm*	*Buckling Class y-y*
ISA 5050(3)@ 2.3 kg/m	*294.862*	*9.70*	*c*
Eff Length m	*Sl Ratio*	*D.C.S.* f_{cd} *N/mm²*	*F.A.L. kN*
1.000	103.093	103.165	30.420
1.100	113.402	90.892	26.801
1.200	123.711	80.211	23.651
1.300	134.021	71.043	20.948
1.400	144.330	63.169	18.626
1.500	154.639	56.463	16.649
1.600	164.948	50.727	14.957
1.700	175.258	45.734	13.485
1.800	185.567	41.429	12.216
1.900	195.876	37.702	11.117
2.000	206.186	34.444	10.156
2.100	216.495	31.546	9.302
2.200	226.804	29.035	8.561
2.300	237.113	26.806	7.904
2.400	247.423	24.790	7.310
2.425	250.000	24.300	7.165

ble X (22) Axial Compression Load Capacity of Single Equal Angle Section

ISA 5050(4)@3.0 kg/m

Section	*Area mm²*	$r_{min} = r_{vv}$ *mm*	*Buckling Class y-y*
ISA 5050(4)@3.0 kg/m	*387.862*	*9.70*	*c*
Eff Length m	*Sl Ratio*	*D.C.S.* f_{cd} *N/mm²*	*F.A.L. kN*
1.000	103.093	103.165	40.014
1.100	113.402	90.892	35.254
1.200	123.711	80.211	31.111
1.300	134.021	71.043	27.555
1.400	144.330	63.169	24.501
1.500	154.639	56.463	21.900
1.600	164.948	50.727	19.675
1.700	175.258	45.734	17.739
1.800	185.567	41.429	16.069
1.900	195.876	37.702	14.623
2.000	206.186	34.444	13.360
2.100	216.495	31.546	12.236
2.200	226.804	29.035	11.262
2.300	237.113	26.806	10.397
2.400	247.423	24.790	9.615
2.425	250.000	24.300	9.425

Table X (23) Axial Compression Load Capacity of Single Equal Angle Section

ISA 5050(5)@3.8 kg/m

Section	*Area mm²*	*$r_{min} = r_{vv}$ mm*	*Buckling Class y-y*
ISA 5050(5)@3.8 kg/m	*478.862*	*9.70*	*c*
Eff Length m	*Sl Ratio*	*D.C.S. f_{cd} N/mm²*	*F.A.L. kN*
1.000	103.093	103.165	49.402
1.100	113.402	90.892	43.525
1.200	123.711	80.211	38.410
1.300	134.021	71.043	34.020
1.400	144.330	63.169	30.249
1.500	154.639	56.463	27.038
1.600	164.948	50.727	24.291
1.700	175.258	45.734	21.900
1.800	185.567	41.429	19.839
1.900	195.876	37.702	18.054
2.000	206.186	34.444	16.494
2.100	216.495	31.546	15.106
2.200	226.804	29.035	13.904
2.300	237.113	26.806	12.836
2.400	247.423	24.790	11.871
2.425	250.000	24.300	11.636

Table X (24) Axial Compression Load Capacity of Single Equal Angle Section

ISA 5050(6)@ 4.5 kg/m

Section	*Area mm²*	*$r_{min} = r_{vv}$ mm*	*Buckling Class y-y*
ISA 5050(6)@ 4.5 kg/m	*567.862*	*9.60*	*c*
Eff Length m	*Sl Ratio*	*D.C.S. f_{cd} N/mm²*	*F.A.L. kN*
1.000	104.167	101.833	57.827
1.100	114.583	89.604	50.883
1.200	125.000	79.000	44.861
1.300	135.417	69.913	39.701
1.400	145.833	62.117	35.274
1.500	156.250	55.513	31.523
1.600	166.667	49.833	28.298
1.700	177.083	44.913	25.504
1.800	187.500	40.675	23.098
1.900	197.917	37.008	21.016
2.000	208.333	33.800	19.194
2.100	218.750	30.938	17.568
2.200	229.167	28.492	16.179
2.300	239.583	26.288	14.928
2.400	250.000	24.300	13.799

able X (25) Axial Compression Load Capacity of Single Equal Angle Section

ISA 5555(5)@4.1 kg/m

Section	Area mm^2	$r_{min} = r_{vv}$ mm	Buckling Class y-y
ISA 5555(5)@4.1 kg/m	527.199	10.60	c
Eff Length m	Sl Ratio	D.C.S. f_{cd} N/mm^2	F.A.L. kN
1.000	94.340	114.925	60.588
1.100	103.774	102.321	53.943
1.200	113.208	91.104	48.030
1.300	122.642	81.217	42.818
1.400	132.075	72.619	38.285
1.500	141.509	65.143	34.344
1.600	150.943	58.643	30.917
1.700	160.377	53.104	27.996
1.800	169.811	48.198	25.410
1.900	179.245	43.940	23.165
2.000	188.679	40.215	21.201
2.100	198.113	36.942	19.476
2.200	207.547	34.036	17.944
2.300	216.981	31.415	16.562
2.400	226.415	29.125	15.354
2.500	235.849	27.072	14.272
2.600	245.283	25.196	13.283
2.650	250.000	24.300	12.811

able X (26) Axial Compression Load Capacity of Single Equal Angle Section

ISA 5555(6)@ 4.9 kg/m

Section	Area mm^2	$r_{min} = r_{vv}$ mm	Buckling Class y-y
ISA 5555(6)@ 4.9 kg/m	626.199	10.60	c
Eff Length m	Sl Ratio	D.C.S. f_{cd} N/mm^2	F.A.L. kN
1.000	94.340	114.925	71.966
1.100	103.774	102.321	64.073
1.200	113.208	91.104	57.049
1.300	122.642	81.217	50.858
1.400	132.075	72.619	45.474
1.500	141.509	65.143	40.793
1.600	150.943	58.643	36.722
1.700	160.377	53.104	33.254
1.800	169.811	48.198	30.182
1.900	179.245	43.940	27.515
2.000	188.679	40.215	25.183
2.100	198.113	36.942	23.133
2.200	207.547	34.036	21.313
2.300	216.981	31.415	19.672
2.400	226.415	29.125	18.238

Contd...

2.500	235.849	27.072	16.952
2.600	245.283	25.196	15.778
2.650	250.000	24.300	15.217

Table X (27) Axial Compression Load Capacity of Single Equal Angle Section

ISA 5555(8)@ 6.4 kg/m

Section	*Area mm²*	$r_{min} = r_{vv}$ *mm*	*Buckling Class y-y*
ISA 5555(8)@ 6.4 kg/m	*818.199*	*10.60*	*c*
Eff Length m	*Sl Ratio*	*D.C.S.* f_{cd} *N/mm²*	*F.A.L. kN*
1.000	94.340	114.925	94.031
1.100	103.774	102.321	83.719
1.200	113.208	91.104	74.541
1.300	122.642	81.217	66.452
1.400	132.075	72.619	59.417
1.500	141.509	65.143	53.300
1.600	150.943	58.643	47.982
1.700	160.377	53.104	43.449
1.800	169.811	48.198	39.436
1.900	179.245	43.940	35.951
2.000	188.679	40.215	32.904
2.100	198.113	36.942	30.226
2.200	207.547	34.036	27.848
2.300	216.981	31.415	25.704
2.400	226.415	29.125	23.830
2.500	235.849	27.072	22.150
2.600	245.283	25.196	20.616
2.650	250.000	24.300	19.882

Table X (28) Axial Compression Load Capacity of Single Equal Angle Section

ISA 5555(10)@7.9 kg/m

Section	*Area mm²*	$r_{min} = r_{vv}$ *mm*	*Buckling Class y-y*
ISA 5555(10)@7.9 kg/m	*1002.199*	*10.60*	*c*
Eff Length m	*Sl Ratio*	*D.C.S.* f_{cd} *N/mm²*	*F.A.L. kN*
1.000	94.340	114.925	115.177
1.100	103.774	102.321	102.546
1.200	113.208	91.104	91.304
1.300	122.642	81.217	81.396
1.400	132.075	72.619	72.779
1.500	141.509	65.143	65.287
1.600	150.943	58.643	58.772
1.700	160.377	53.104	53.221
1.800	169.811	48.198	48.304
1.900	179.245	43.940	44.036
2.000	188.679	40.215	40.304

Contd...

2.100	198.113	36.942	37.023
2.200	207.547	34.036	34.111
2.300	216.981	31.415	31.484
2.400	226.415	29.125	29.189
2.500	235.849	27.072	27.131
2.600	245.283	25.196	25.252
2.650	250.000	24.300	24.353

Table X (29) Axial Compression Load Capacity of Single Equal Angle Section

ISA6060(5)@ 4.5 kg/m

Section	*Area mm²*	*$r_{min} = r_{vv}$ mm*	*Buckling Class y-y*
ISA6060(5)@ 4.5 kg/m	*575.375*	*11.60*	*c*
Eff Length m	*Sl Ratio*	*D.C.S. f_{cd} N/mm²*	*F.A.L. kN*
1.000	86.207	126.690	72.894
1.100	94.828	114.241	65.732
1.200	103.448	102.724	59.105
1.300	112.069	92.345	53.133
1.400	120.690	83.052	47.786
1.500	129.310	74.948	43.123
1.600	137.931	67.876	39.054
1.700	146.552	61.614	35.451
1.800	155.172	56.148	32.306
1.900	163.793	51.328	29.533
2.000	172.414	47.014	27.051
2.100	181.034	43.197	24.854
2.200	189.655	39.834	22.920
2.300	198.276	36.886	21.223
2.400	206.897	34.231	19.696
2.500	215.517	31.810	18.303
2.600	224.138	29.648	17.059
2.700	232.759	27.721	15.950
2.800	241.379	25.938	14.924
2.900	250.000	24.300	13.982

Table X (30) Axial Compression Load Capacity of Single Equal Angle Section

ISA 6060(6)@ 5.4 kg/m

Section	*Area mm²*	*$r_{min} = r_{vv}$ mm*	*Buckling Class y-y*
ISA 6060(6)@ 5.4 kg/m	*634.375*	*11.50*	*c*
Eff Length m	*Sl Ratio*	*D.C.S. f_{cd} N/mm²*	*F.A.L. kN*
1.000	86.957	125.565	85.934
1.100	95.652	113.087	77.394
1.200	104.348	101.609	69.539
1.300	113.043	91.283	62.472
1.400	121.739	82.065	56.163

Contd...

1.500	130.435	73.948	50.608
1.600	139.130	66.904	45.788
1.700	147.826	60.722	41.556
1.800	156.522	55.352	37.882
1.900	165.217	50.587	34.620
2.000	173.913	46.339	31.713
2.100	182.609	42.583	29.142
2.200	191.304	39.257	26.866
2.300	200.000	36.300	24.843
2.400	208.696	33.691	23.058
2.500	217.391	31.304	21.424
2.600	226.087	29.200	19.984
2.700	234.783	27.296	18.680
2.800	243.478	25.539	17.478
2.875	250.000	24.300	16.630

Table X (31) Axial Compression Load Capacity of Single Equal Angle Section

ISA 6060(8)@ 7.0 kg/m

Section	*Area mm²*	$r_{min} = r_{vv}$ *mm*	*Buckling Class y-y*
ISA 6060(8)@ 7.0 kg/m	*896.375*	*11.50*	*c*
Eff Length m	*Sl Ratio*	*D.C.S.* f_{cd} *N/mm²*	*F.A.L. kN*
1.000	86.957	125.565	112.554
1.100	95.652	113.087	101.368
1.200	104.348	101.609	91.080
1.300	113.043	91.283	81.823
1.400	121.739	82.065	73.561
1.500	130.435	73.948	66.285
1.600	139.130	66.904	59.971
1.700	147.826	60.722	54.429
1.800	156.522	55.352	49.616
1.900	165.217	50.587	45.345
2.000	173.913	46.339	41.537
2.100	182.609	42.583	38.170
2.200	191.304	39.257	35.189
2.300	200.000	36.300	32.538
2.400	208.696	33.691	30.200
2.500	217.391	31.304	28.060
2.600	226.087	29.200	26.174
2.700	234.783	27.296	24.467
2.800	243.478	25.539	22.893
2.875	250.000	24.300	21.782

Table X (32) Axial Compression Load Capacity of Single Equal Angle Section

ISA 6060(10)@ 8.6 kg/m

Section	Area mm²	$r_{min} = r_{vv}$ mm	Buckling Class y-y
ISA 6060(10)@ 8.6 kg/m	1100.375	11.50	c
Eff Length m	Sl Ratio	D.C.S. f_{cd} N/mm²	F.A.L. kN
1.000	86.957	125.565	138.169
1.100	95.652	113.087	124.438
1.200	104.348	101.609	111.808
1.300	113.043	91.283	100.445
1.400	121.739	82.065	90.303
1.500	130.435	73.948	81.370
1.600	139.130	66.904	73.620
1.700	147.826	60.722	66.817
1.800	156.522	55.352	60.908
1.900	165.217	50.587	55.665
2.000	173.913	46.339	50.990
2.100	182.609	42.583	46.857
2.200	191.304	39.257	43.197
2.300	200.000	36.300	39.944
2.400	208.696	33.691	37.073
2.500	217.391	31.304	34.447
2.600	226.087	29.200	32.131
2.700	234.783	27.296	30.035
2.800	243.478	25.539	28.103
2.875	250.000	24.300	26.739

Table X (33) Axial Compression Load Capacity of Single Equal Angle Section

ISA 6565(5)@ 4.9 kg/m

Section	Area mm²	$r_{min} = r_{vv}$ mm	Buckling Class y-y
ISA 6565(5)@ 4.9 kg/m	625.375	12.60	c
Eff Length m	Sl Ratio	D.C.S. f_{cd} N/mm²	F.A.L. kN
1.000	79.365	137.016	85.686
1.100	87.302	125.048	78.202
1.200	95.238	113.667	71.084
1.300	103.175	103.063	64.453
1.400	111.111	93.389	58.403
1.500	119.048	84.738	52.993
1.600	126.984	77.135	48.238
1.700	134.921	70.314	43.973
1.800	142.857	64.200	40.149
1.900	150.794	58.732	36.729
2.000	158.730	54.049	33.801
2.100	166.667	49.833	31.165
2.200	174.603	46.029	28.785

Contd...

2.300	182.540	42.610	26.647
2.400	190.476	39.538	24.726
2.500	198.413	36.840	23.039
2.600	206.349	34.395	21.510
2.700	214.286	32.143	20.101
2.800	222.222	30.089	18.817
2.900	230.159	28.267	17.677
3.000	238.095	26.600	16.635
3.100	246.032	25.054	15.668
3.150	250.000	24.300	15.197

Table X (34) Axial Compression Load Capacity of Single Equal Angle Section

ISA 6565(6)@ 5.8 kg/m

Section	*Area mm²*	*$r_{min} = r_{vv}$ mm*	*Buckling Class y-y*
ISA 6565(6)@ 5.8 kg/m	*744.375*	*12.60*	*c*
Eff Length m	*Sl Ratio*	*D.C.S. f_{cd} N/mm²*	*F.A.L. kN*
1.000	79.365	137.016	101.991
1.100	87.302	125.048	93.082
1.200	95.238	113.667	84.611
1.300	103.175	103.063	76.718
1.400	111.111	93.389	69.516
1.500	119.048	84.738	63.077
1.600	126.984	77.135	57.417
1.700	134.921	70.314	52.340
1.800	142.857	64.200	47.789
1.900	150.794	58.732	43.718
2.000	158.730	54.049	40.233
2.100	166.667	49.833	37.095
2.200	174.603	46.029	34.263
2.300	182.540	42.610	31.717
2.400	190.476	39.538	29.431
2.500	198.413	36.840	27.423
2.600	206.349	34.395	25.603
2.700	214.286	32.143	23.926
2.800	222.222	30.089	22.397
2.900	230.159	28.267	21.041
3.000	238.095	26.600	19.800
3.100	246.032	25.054	18.650
3.150	250.000	24.300	18.088

Table X (35) Axial Compression Load Capacity of Single Equal Angle Section

ISA 6565(8)@ 7.7 kg/m

Section	Area mm^2	$r_{min} = r_{vv}$ mm	Buckling Class y-y
ISA 6565(8)@ 7.7 kg/m	976.375	12.50	c
Eff Length m	Sl Ratio	D.C.S. f_{cd} N/mm^2	F.A.L. kN
1.000	80.000	136.000	132.787
1.100	88.000	124.000	121.071
1.200	96.000	112.600	109.940
1.300	104.000	102.040	99.629
1.400	112.000	92.420	90.237
1.500	120.000	83.700	81.723
1.600	128.000	76.180	74.380
1.700	136.000	69.440	67.800
1.800	144.000	63.400	61.902
1.900	152.000	58.020	56.649
2.000	160.000	53.300	52.041
2.100	168.000	49.140	47.979
2.200	176.000	45.400	44.327
2.300	184.000	42.040	41.047
2.400	192.000	39.020	38.098
2.500	200.000	36.300	35.442
2.600	208.000	33.900	33.099
2.700	216.000	31.680	30.932
2.800	224.000	29.680	28.979
2.900	232.000	27.880	27.221
3.000	240.000	26.200	25.581
3.100	248.000	24.680	24.097
3.125	250.000	24.300	23.726

Table X (36) Axial Compression Load Capacity of Single Equal Angle Section

ISA 6565(10)@ 9.4 kg/m

Section	Area mm^2	$r_{min} = r_{vv}$ mm	Buckling Class y-y
ISA 6565(10)@ 9.4 kg/m	1200.375	12.50	c
Eff Length m	Sl Ratio	D.C.S. f_{cd} N/mm^2	F.A.L. kN
1.000	80.000	136.000	163.251
1.100	88.000	124.000	148.847
1.200	96.000	112.600	135.162
1.300	104.000	102.040	122.486
1.400	112.000	92.420	110.939
1.500	120.000	83.700	100.471
1.600	128.000	76.180	91.445
1.700	136.000	69.440	83.354
1.800	144.000	63.400	76.104
1.900	152.000	58.020	69.646

Contd...

2.000	160.000	53.300	63.980
2.100	168.000	49.140	58.986
2.200	176.000	45.400	54.497
2.300	184.000	42.040	50.464
2.400	192.000	39.020	46.839
2.500	200.000	36.300	43.574
2.600	208.000	33.900	40.693
2.700	216.000	31.680	38.028
2.800	224.000	29.680	35.627
2.900	232.000	27.880	33.466
3.000	240.000	26.200	31.450
3.100	248.000	24.680	29.625
3.125	250.000	24.300	29.169

Table X (37) Axial Compression Load Capacity of Single Equal Angle Section

ISA 7070(5)@ 5.3 kg/m

Section	*Area mm²*	*$r_{min} = r_{vv}$ mm*	*Buckling Class y-y*
ISA 7070(5)@ 5.3 kg/m	*676.824*	*13.60*	*c*
Eff Length m	*Sl Ratio*	*D.C.S. f_{cd} N/mm²*	*F.A.L. kN*
1.000	73.529	146.353	99.055
1.100	80.882	134.676	91.152
1.200	88.235	123.647	83.687
1.300	95.588	113.176	76.601
1.400	102.941	103.353	69.952
1.500	110.294	94.279	63.811
1.600	117.647	86.265	58.386
1.700	125.000	79.000	53.469
1.800	132.353	72.394	48.998
1.900	139.706	66.438	44.967
2.000	147.059	61.259	41.461
2.100	154.412	56.597	38.306
2.200	161.765	52.382	35.454
2.300	169.118	48.559	32.866
2.400	176.471	45.188	30.584
2.500	183.824	42.109	28.500
2.600	191.176	39.300	26.599
2.700	198.529	36.800	24.907
2.800	205.882	34.535	23.374
2.900	213.235	32.426	21.947
3.000	220.588	30.465	20.619
3.100	227.941	28.774	19.475
3.200	235.294	27.188	18.402
3.300	242.647	25.697	17.392
3.400	250.000	24.300	16.447

able X (38) Axial Compression Load Capacity of Single Equal Angle Section

ISA 7070(6)@ 6.3 kg/m

Section	Area mm²	$r_{min} = r_{vv}$ mm	Buckling Class y-y
ISA 7070(6)@ 6.3 kg/m	805.824	13.60	c
Eff Length m	Sl Ratio	D.C.S. f_{cd} N/mm²	F.A.L. kN
1.000	73.529	146.353	117.935
1.100	80.882	134.676	108.526
1.200	88.235	123.647	99.638
1.300	95.588	113.176	91.200
1.400	102.941	103.353	83.284
1.500	110.294	94.279	75.973
1.600	117.647	86.265	69.514
1.700	125.000	79.000	63.660
1.800	132.353	72.394	58.337
1.900	139.706	66.438	53.538
2.000	147.059	61.259	49.364
2.100	154.412	56.597	45.607
2.200	161.765	52.382	42.211
2.300	169.118	48.559	39.130
2.400	176.471	45.188	36.414
2.500	183.824	42.109	33.932
2.600	191.176	39.300	31.669
2.700	198.529	36.800	29.654
2.800	205.882	34.535	27.829
2.900	213.235	32.426	26.130
3.000	220.588	30.465	24.549
3.100	227.941	28.774	23.186
3.200	235.294	27.188	21.909
3.300	242.647	25.697	20.707
3.400	250.000	24.300	19.582

able X (39) Axial Compression Load Capacity of Single Equal Angle Section

ISA 7070(8)@ 8.3 kg/m

Section	Area mm²	$r_{min} = r_{vv}$ mm	Buckling Class y-y
ISA 7070(8)@ 8.3 kg/m	1057.824	13.50	c
Eff Length m	Sl Ratio	D.C.S. f_{cd} N/mm²	F.A.L. kN
1.000	74.074	145.481	153.894
1.100	81.481	133.778	141.513
1.200	88.889	122.667	129.760
1.300	96.296	112.185	118.672
1.400	103.704	102.407	108.329
1.500	111.111	93.389	98.789
1.600	118.519	85.315	90.248
1.700	125.926	78.130	82.647

Contd...

1.800	133.333	71.600	75.740
1.900	140.741	65.681	69.479
2.000	148.148	60.496	63.994
2.100	155.556	55.922	59.156
2.200	162.963	51.759	54.752
2.300	170.370	47.933	50.705
2.400	177.778	44.600	47.179
2.500	185.185	41.578	43.982
2.600	192.593	38.819	41.063
2.700	200.000	36.300	38.399
2.800	207.407	34.078	36.048
2.900	214.815	32.000	33.850
3.000	222.222	30.089	31.829
3.100	229.630	28.385	30.027
3.200	237.037	26.822	28.373
3.300	244.444	25.356	26.822
3.375	250.000	24.300	25.705

Table X (40) Axial Compression Load Capacity of Single Equal Angle Section

ISA 7070(10)@ 10.2 kg/m

Section	*Area mm²*	*$r_{min} = r_{vv}$ mm*	*Buckling Class y-y*
ISA 7070(10)@ 10.2 kg/m	*1301.824*	*13.50*	*c*
Eff Length m	*Sl Ratio*	*D.C.S. f_{cd} N/mm²*	*F.A.L. kN*
1.000	74.074	145.481	189.391
1.100	81.481	133.778	174.155
1.200	88.889	122.667	159.690
1.300	96.296	112.185	146.045
1.400	103.704	102.407	133.316
1.500	111.111	93.389	121.576
1.600	118.519	85.315	111.065
1.700	125.926	78.130	101.711
1.800	133.333	71.600	93.211
1.900	140.741	65.681	85.506
2.000	148.148	60.496	78.756
2.100	155.556	55.922	72.801
2.200	162.963	51.759	67.381
2.300	170.370	47.933	62.401
2.400	177.778	44.600	58.061
2.500	185.185	41.578	54.127
2.600	192.593	38.819	50.535
2.700	200.000	36.300	47.256
2.800	207.407	34.078	44.363
2.900	214.815	32.000	41.658
3.000	222.222	30.089	39.170
3.100	229.630	28.385	36.953

Contd...

3.200	237.037	26.822	34.918
3.300	244.444	25.356	33.008
3.375	250.000	24.300	31.634

able X (41) Axial Compression Load Capacity of Single Equal Angle Section

ISA 7575(5)@ 5.7 kg/m

Section	*Area mm²*	*$r_{min} = r_{vv}$ mm*	*Buckling Class y-y*
ISA 7575(5)@ 5.7 kg/m	*726.824*	*14.60*	*c*
Eff Length m	*Sl Ratio*	*D.C.S. f_{cd} N/mm²*	*F.A.L. kN*
1.000	68.493	154.411	112.230
1.200	82.192	132.712	96.459
1.400	95.890	112.753	81.952
1.600	109.589	95.110	69.128
1.800	123.288	80.610	58.589
2.000	136.986	68.641	49.890
2.200	150.685	58.796	42.734
2.400	164.384	51.021	37.083
2.600	178.082	44.463	32.317
2.800	191.781	39.095	28.415
3.000	205.479	34.656	25.189
3.200	219.178	30.822	22.402
3.400	232.877	27.696	20.130
3.600	246.575	24.819	18.039
3.650	250.000	24.300	17.662

able X (42) Axial Compression Load Capacity of Single Equal Angle Section

ISA 7575(6)@6.8 kg/m

Section	*Area mm²*	*$r_{min} = r_{vv}$ mm*	*Buckling Class y-y*
ISA 7575(6)@6.8 kg/m	*865.824*	*14.60*	*c*
Eff Length m	*Sl Ratio*	*D.C.S. f_{cd} N/mm²*	*F.A.L. kN*
1.000	68.493	154.411	133.693
1.200	82.192	132.712	114.906
1.400	95.890	112.753	97.625
1.600	109.589	95.110	82.348
1.800	123.288	80.610	69.794
2.000	136.986	68.641	59.431
2.200	150.685	58.796	50.907
2.400	164.384	51.021	44.175
2.600	178.082	44.463	38.497
2.800	191.781	39.095	33.849
3.000	205.479	34.656	30.006
3.200	219.178	30.822	26.686
3.400	232.877	27.696	23.980
3.600	246.575	24.819	21.489
3.650	250.000	24.300	21.040

Table X (43) Axial Compression Load Capacity Of Single Equal Angle Section

ISA 7575(8)@ 8.9 kg/m

Section	Area mm²	$r_{min} = r_{vv}$ mm	Buckling Class y-y
ISA 7575(8)@ 8.9 kg/m	1137.824	14.50	c
Eff Length m	Sl Ratio	D.C.S. f_{cd} N/mm²	F.A.L. kN
1.000	68.966	153.655	174.833
1.200	82.759	131.862	150.036
1.400	96.552	111.828	127.240
1.600	110.345	94.224	107.210
1.800	124.138	79.810	90.810
2.000	137.931	67.876	77.231
2.200	151.724	58.183	66.202
2.400	165.517	50.431	57.382
2.600	179.310	43.910	49.962
2.800	193.103	38.645	43.971
3.000	206.897	34.231	38.949
3.200	220.690	30.441	34.637
3.400	234.483	27.359	31.129
3.600	248.276	24.628	28.022
3.625	250.000	24.300	27.649

Table X (44) Axial Compression Load Capacity of Single Equal Angle Section

ISA 7575(10)@ 11.0 kg/m

Section	Area mm²	$r_{min} = r_{vv}$ mm	Buckling Class y-y
ISA 7575(10)@ 11.0 kg/m	1401.824	14.50	c
Eff Length m	Sl Ratio	D.C.S. f_{cd} N/mm²	F.A.L. kN
1.000	68.966	153.655	215.398
1.200	82.759	131.862	184.847
1.400	96.552	111.828	156.763
1.600	110.345	94.224	132.086
1.800	124.138	79.810	111.880
2.000	137.931	67.876	95.150
2.200	151.724	58.183	81.562
2.400	165.517	50.431	70.695
2.600	179.310	43.910	61.555
2.800	193.103	38.645	54.173
3.000	206.897	34.231	47.986
3.200	220.690	30.441	42.673
3.400	234.483	27.359	38.352
3.600	248.276	24.628	34.524
3.625	250.000	24.300	34.064

Table X (45) Axial Compression Load Capacity of Single Equal Angle Section

ISA 8080(6)@ 7.3 kg/m

Section	Area mm^2	$r_{min} = r_{vv}$ mm	Buckling Class y-y
ISA 8080(6)@ 7.3 kg/m	929.043	15.60	c
Eff Length m	Sl Ratio	D.C.S. f_{cd} N/mm^2	F.A.L. kN
1.000	64.103	161.436	149.981
1.200	76.923	140.923	130.924
1.400	89.744	121.385	112.772
1.600	102.564	103.821	96.454
1.800	115.385	88.731	82.435
2.000	128.205	75.987	70.595
2.200	141.026	65.482	60.836
2.400	153.846	56.931	52.891
2.600	166.667	49.833	46.297
2.800	179.487	43.831	40.721
3.000	192.308	38.915	36.154
3.200	205.128	34.762	32.295
3.400	217.949	31.154	28.943
3.600	230.769	28.138	26.142
3.800	243.590	25.518	23.707
3.900	250.000	24.300	22.576

Table X (46) Axial Compression Load Capacity of Single Equal Angle Section

ISA 8080(8)@ 9.6 kg/m

Section	Area mm^2	$r_{min} = r_{vv}$ mm	Buckling Class y-y
ISA 8080(8)@ 9.6 kg/m	1221.043	15.50	c
Eff Length m	Sl Ratio	D.C.S. f_{cd} N/mm^2	F.A.L. kN
1.000	64.516	160.774	196.312
1.200	77.419	140.129	171.104
1.400	90.323	120.548	147.195
1.600	103.226	103.000	125.767
1.800	116.129	87.919	107.353
2.000	129.032	75.210	91.834
2.200	141.935	64.845	79.179
2.400	154.839	56.345	68.800
2.600	167.742	49.274	60.166
2.800	180.645	43.348	52.930
3.000	193.548	38.494	47.002
3.200	206.452	34.365	41.961
3.400	219.355	30.774	37.577
3.600	232.258	27.826	33.977
3.800	245.161	25.219	30.794
3.875	250.000	24.300	29.671

Table X (47) Axial Compression Load Capacity of Single Equal Angle Section

ISA 8080(10)@ 11.8 kg/m

Section	Area mm^2	$r_{min} = r_{vv}$ mm	Buckling Class y-y
ISA 8080(10)@ 11.8 kg/m	1505.043	15.50	c
Eff Length m	Sl Ratio	D.C.S. f_{cd} N/mm^2	F.A.L. kN
1.000	64.516	160.774	241.972
1.200	77.419	140.129	210.900
1.400	90.323	120.548	181.431
1.600	103.226	103.000	155.019
1.800	116.129	87.919	132.322
2.000	129.032	75.210	113.194
2.200	141.935	64.845	97.595
2.400	154.839	56.345	84.802
2.600	167.742	49.274	74.160
2.800	180.645	43.348	65.241
3.000	193.548	38.494	57.934
3.200	206.452	34.365	51.720
3.400	219.355	30.774	46.316
3.600	232.258	27.826	41.879
3.800	245.161	25.219	37.956
3.875	250.000	24.300	36.573

Table X (48) Axial Compression Load Capacity of Single Equal Angle Section

ISA 8080(12)@ 14.0 kg/m

Section	Area mm^2	$r_{min} = r_{vv}$ mm	Buckling Class y-y
ISA 8080(12)@ 14.0 kg/m	1781.043	15.40	c
Eff Length m	Sl Ratio	D.C.S. f_{cd} N/mm^2	F.A.L. kN
1.000	64.935	160.104	285.152
1.200	77.922	139.325	248.143
1.400	90.909	119.727	213.239
1.600	103.896	102.169	181.967
1.800	116.883	87.097	155.124
2.000	129.870	74.422	132.549
2.200	142.857	64.200	114.343
2.400	155.844	55.752	99.297
2.600	168.831	48.708	86.751
2.800	181.818	42.891	76.391
3.000	194.805	38.066	67.798
3.200	207.792	33.962	60.488
3.400	220.779	30.421	54.181
3.600	233.766	27.509	48.995
3.800	246.753	24.917	44.378
3.850	250.000	24.300	43.279

Table X (49) Axial Compression Load Capacity of Single Equal Angle Section

ISA 9090(6)@ 8.2 kg/m

Section	Area mm^2	$r_{min} = r_{vv}$ mm	Buckling Class y-y
ISA 9090(6)@ 8.2 kg/m	1046.521	17.50	c
Eff Length m	Sl Ratio	D.C.S. f_{cd} N/mm^2	F.A.L. kN
1.000	57.143	172.286	180.301
1.200	68.571	154.286	161.463
1.400	80.000	136.000	142.327
1.600	91.429	119.000	124.536
1.800	102.857	103.457	108.270
2.000	114.286	89.929	94.112
2.200	125.714	78.329	81.973
2.400	137.143	68.514	71.702
2.600	148.571	60.200	63.001
2.800	160.000	53.300	55.780
3.000	171.429	47.457	49.665
3.200	182.857	42.486	44.462
3.400	194.286	38.243	40.022
3.600	205.714	34.586	36.195
3.800	217.143	31.371	32.831
4.000	228.571	28.629	29.960
4.200	240.000	26.200	27.419
4.375	250.000	24.300	25.430

Table X (50) Axial Compression Load Capacity of Single Equal Angle Section

ISA 9090(8)@ 10.8 kg/m

Section	Area mm^2	$r_{min} = r_{vv}$ mm	Buckling Class y-y
ISA 9090(8)@ 10.8 kg/m	1378.521	17.50	c
Eff Length m	Sl Ratio	D.C.S. f_{cd} N/mm^2	F.A.L. kN
1.000	57.143	172.286	237.500
1.200	68.571	154.286	212.686
1.400	80.000	136.000	187.479
1.600	91.429	119.000	164.044
1.800	102.857	103.457	142.618
2.000	114.286	89.929	123.968
2.200	125.714	78.329	107.978
2.400	137.143	68.514	94.448
2.600	148.571	60.200	82.987
2.800	160.000	53.300	73.475
3.000	171.429	47.457	65.421
3.200	182.857	42.486	58.567
3.400	194.286	38.243	52.719
3.600	205.714	34.586	47.677
3.800	217.143	31.371	43.246

Contd...

4.000	228.571	28.629	39.465
4.200	240.000	26.200	36.117
4.375	250.000	24.300	33.498

Table X (51) Axial Compression Load Capacity of Single Equal Angle Section

ISA 9090(10)@ 13.4 kg/m

Section	*Area mm²*	$r_{min} = r_{vv}$ *mm*	*Buckling Class y-y*
ISA 9090(10)@ 13.4 kg/m	*1702.521*	*17.40*	*c*
Eff Length m	*Sl Ratio*	*D.C.S.* f_{cd} *N/mm²*	*F.A.L. kN*
1.000	57.471	171.793	292.481
1.200	68.966	153.655	261.601
1.400	80.460	135.310	230.369
1.600	91.954	118.264	201.348
1.800	103.448	102.724	174.890
2.000	114.943	89.213	151.886
2.200	126.437	77.649	132.200
2.400	137.931	67.876	115.560
2.600	149.425	59.602	101.474
2.800	160.920	52.822	89.930
3.000	172.414	47.014	80.042
3.200	183.908	42.076	71.635
3.400	195.402	37.863	64.463
3.600	206.897	34.231	58.279
3.800	218.391	31.034	52.837
4.000	229.885	28.326	48.226
4.200	241.379	25.938	44.160
4.350	250.000	24.300	41.371

Table X (52) Axial Compression Load Capacity of Single Equal Angle Section

ISA 9090(12)@ 15.8 kg/m

Section	*Area mm²*	$r_{min} = r_{vv}$ *mm*	*Buckling Class y-y*
ISA 9090(12)@ 15.8 kg/m	*2018.521*	*17.40*	*c*
Eff Length m	*Sl Ratio*	*D.C.S.* f_{cd} *N/mm²*	*F.A.L. kN*
1.000	57.471	171.793	346.768
1.200	68.966	153.655	310.156
1.400	80.460	135.310	273.127
1.600	91.954	118.264	238.719
1.800	103.448	102.724	207.351
2.000	114.943	89.213	180.078
2.200	126.437	77.649	156.737
2.400	137.931	67.876	137.009
2.600	149.425	59.602	120.309
2.800	160.920	52.822	106.622
3.000	172.414	47.014	94.898

Contd..

3.200	183.908	42.076	84.931
3.400	195.402	37.863	76.428
3.600	206.897	34.231	69.096
3.800	218.391	31.034	62.644
4.000	229.885	28.326	57.178
4.200	241.379	25.938	52.356
4.350	250.000	24.300	49.050

able X (53) Axial Compression Load Capacity of Single Equal Angle Section

ISA 100100(6)@ 9.2 kg/m

Section	*Area mm²*	$r_{min} = r_{vv}$ *mm*	*Buckling Class y-y*
ISA 100100(6)@ 9.2 kg/m	*1166.521*	*19.50*	*c*
Eff Length m	*Sl Ratio*	*D.C.S.* f_{cd} *N/mm²*	*F.A.L. kN*
1.000	51.282	181.077	211.230
1.200	61.538	165.538	193.104
1.400	71.795	149.128	173.961
1.600	82.051	132.923	155.058
1.800	92.308	117.769	137.380
2.000	102.564	103.821	121.109
2.200	112.821	91.526	106.767
2.400	123.077	80.808	94.264
2.600	133.333	71.600	83.523
2.800	143.590	63.687	74.292
3.000	153.846	56.931	66.411
3.200	164.103	51.167	59.687
3.400	174.359	46.138	53.822
3.600	184.615	41.800	48.761
3.800	194.872	38.044	44.379
4.000	205.128	34.762	40.550
4.200	215.385	31.846	37.149
4.400	225.641	29.303	34.182
4.600	235.897	27.062	31.568
4.800	246.154	25.031	29.199
4.875	250.000	24.300	28.346

able X (54) Axial Compression Load Capacity of Single Equal Angle Section

ISA 100100(8)@ 12.1 kg/m

Section	*Area mm²*	$r_{min} = r_{vv}$ *mm*	*Buckling Class y-y*
ISA 100100(8)@ 12.1 kg/m	*1538.521*	*19.50*	*c*
Eff Length m	*Sl Ratio*	*D.C.S.* f_{cd} *N/mm²*	*F.A.L. kN*
1.000	51.282	181.077	278.591
1.200	61.538	165.538	254.684
1.400	71.795	149.128	229.437
1.600	82.051	132.923	204.505
1.800	92.308	117.769	181.191

Contd...

2.000	102.564	103.821	159.730
2.200	112.821	91.526	140.814
2.400	123.077	80.808	124.324
2.600	133.333	71.600	110.158
2.800	143.590	63.687	97.984
3.000	153.846	56.931	87.589
3.200	164.103	51.167	78.721
3.400	174.359	46.138	70.985
3.600	184.615	41.800	64.310
3.800	194.872	38.044	58.531
4.000	205.128	34.762	53.481
4.200	215.385	31.846	48.996
4.400	225.641	29.303	45.083
4.600	235.897	27.062	41.635
4.800	246.154	25.031	38.510
4.875	250.000	24.300	37.386

Table X (55) Axial Compression Load Capacity Of Single Equal Angle Section

ISA 100100(10)@ 14.9 kg/m

Section	*Area mm²*	$r_{min} = r_{vv}$ *mm*	*Buckling Class y-y*
ISA 100100(10)@ 14.9 kg/m	*1902.521*	*19.40*	*c*
Eff Length m	*Sl Ratio*	*D.C.S.* f_{cd} *N/mm²*	*F.A.L. kN*
1.000	51.546	180.680	343.748
1.200	61.856	165.031	313.975
1.400	72.165	148.536	282.593
1.600	82.474	132.289	251.682
1.800	92.784	117.103	222.791
2.000	103.093	103.165	196.274
2.200	113.402	90.892	172.924
2.400	123.711	80.211	152.604
2.600	134.021	71.043	135.161
2.800	144.330	63.169	120.181
3.000	154.639	56.463	107.422
3.200	164.948	50.727	96.509
3.400	175.258	45.734	87.010
3.600	185.567	41.429	78.819
3.800	195.876	37.702	71.729
4.000	206.186	34.444	65.531
4.200	216.495	31.546	60.018
4.400	226.804	29.035	55.240
4.600	237.113	26.806	50.999
4.800	247.423	24.790	47.163
4.850	250.000	24.300	46.231

able X (56) Axial Compression Load Capacity of Single Equal Angle Section

ISA 100100(12)@ 17.7 kg/m

Section	Area mm^2	$r_{min} = r_{vv}$ mm	Buckling Class y-y
ISA 100100(12)@ 17.7 kg/m	2258.521	19.40	c
Eff Length m	Sl Ratio	D.C.S. f_{cd} N/mm^2	F.A.L. kN
1.000	51.546	180.680	408.071
1.200	61.856	165.031	372.726
1.400	72.165	148.536	335.472
1.600	82.474	132.289	298.777
1.800	92.784	117.103	264.480
2.000	103.093	103.165	233.000
2.200	113.402	90.892	205.281
2.400	123.711	80.211	181.159
2.600	134.021	71.043	160.453
2.800	144.330	63.169	142.669
3.000	154.639	56.463	127.523
3.200	164.948	50.727	114.568
3.400	175.258	45.734	103.291
3.600	185.567	41.429	93.568
3.800	195.876	37.702	85.151
4.000	206.186	34.444	77.793
4.200	216.495	31.546	71.248
4.400	226.804	29.035	65.576
4.600	237.113	26.806	60.542
4.800	247.423	24.790	55.988
4.850	250.000	24.300	54.882

able X (57) Axial Compression Load Capacity of Single Equal Angle Section

ISA 110110(8)@ 13.4 kg/m

Section	Area mm^2	$r_{min} = r_{vv}$ mm	Buckling Class y-y
ISA 110110(8)@ 13.4 kg/m	1702.008	21.40	c
Eff Length m	Sl Ratio	D.C.S. f_{cd} N/mm^2	F.A.L. kN
1.000	46.729	187.907	319.819
1.200	56.075	173.888	295.959
1.400	65.421	159.327	271.176
1.600	74.766	144.374	245.726
1.800	84.112	129.832	220.975
2.000	93.458	116.159	197.703
2.200	102.804	103.523	176.198
2.400	112.150	92.257	157.022
2.600	121.495	82.294	140.066
2.800	130.841	73.619	125.300
3.000	140.187	66.069	112.450
3.200	149.533	59.527	101.316
3.400	158.879	53.962	91.843

Contd...

3.600	168.224	49.023	83.438
3.800	177.570	44.693	76.069
4.000	186.916	40.903	69.617
4.200	196.262	37.571	63.946
4.400	205.607	34.618	58.920
4.600	214.953	31.963	54.401
4.800	224.299	29.611	50.399
5.000	233.645	27.535	46.864
5.200	242.991	25.632	43.626
5.350	250.000	24.300	41.359

Table X (58) Axial Compression Load Capacity of Single Equal Angle Section

ISA 110110(10)@ 16.5 kg/m

Section	*Area mm²*	*$r_{min} = r_{vv}$ mm*	*Buckling Class y-y*
ISA 110110(10)@ 16.5 kg/m	*2106.008*	*21.40*	*c*
Eff Length m	*Sl Ratio*	*D.C.S. f_{cd} N/mm²*	*F.A.L. kN*
1.000	46.729	187.907	395.733
1.200	56.075	173.888	366.209
1.400	65.421	159.327	335.544
1.600	74.766	144.374	304.053
1.800	84.112	129.832	273.427
2.000	93.458	116.159	244.632
2.200	102.804	103.523	218.021
2.400	112.150	92.257	194.294
2.600	121.495	82.294	173.313
2.800	130.841	73.619	155.042
3.000	140.187	66.069	139.142
3.200	149.533	59.527	125.365
3.400	158.879	53.962	113.644
3.600	168.224	49.023	103.244
3.800	177.570	44.693	94.125
4.000	186.916	40.903	86.142
4.200	196.262	37.571	79.125
4.400	205.607	34.618	72.905
4.600	214.953	31.963	67.314
4.800	224.299	29.611	62.361
5.000	233.645	27.535	57.988
5.200	242.991	25.632	53.981
5.350	250.000	24.300	51.176

able X (59) Axial Compression Load Capacity of Single Equal Angle Section

ISA 110110(12)@ 19.6 kg/m

Section	Area mm^2	$r_{min} = r_{vv}$ mm	Buckling Class y-y
ISA 110110(12)@ 19.6 kg/m	2502.008	21.30	c
Eff Length m	Sl Ratio	D.C.S. f_{cd} N/mm^2	F.A.L. kN
1.000	46.948	187.577	469.320
1.200	56.338	173.493	434.081
1.400	65.728	158.836	397.408
1.600	75.117	143.812	359.819
1.800	84.507	129.239	323.358
2.000	93.897	115.545	289.094
2.200	103.286	102.925	257.519
2.400	112.676	91.683	229.392
2.600	122.066	81.758	204.560
2.800	131.455	73.121	182.950
3.000	140.845	65.608	164.153
3.200	150.235	59.062	147.772
3.400	159.624	53.522	133.912
3.600	169.014	48.613	121.629
3.800	178.404	44.318	110.885
4.000	187.793	40.561	101.483
4.200	197.183	37.258	93.219
4.400	206.573	34.328	85.889
4.600	215.962	31.690	79.289
4.800	225.352	29.369	73.482
5.000	234.742	27.304	68.315
5.200	244.131	25.415	63.589
5.325	250.000	24.300	60.799

able X (60) Axial Compression Load Capacity of Single Equal Angle Section

ISA 110110(15)@ 24.2 kg/m

Section	Area mm^2	$r_{min} = r_{vv}$ mm	Buckling Class y-y
ISA 110110(15)@ 24.2 kg/m	3081.008	21.30	c
Eff Length m	Sl Ratio	D.C.S. f_{cd} N/mm^2	F.A.L. kN
1.000	46.948	187.577	577.928
1.200	56.338	173.493	534.533
1.400	65.728	158.836	489.374
1.600	75.117	143.812	443.087
1.800	84.507	129.239	398.188
2.000	93.897	115.545	355.994
2.200	103.286	102.925	317.112
2.400	112.676	91.683	282.476
2.600	122.066	81.758	251.898
2.800	131.455	73.121	225.287
3.000	140.845	65.608	202.140

Contd...

3.200	150.235	59.062	181.969
3.400	159.624	53.522	164.901
3.600	169.014	48.613	149.776
3.800	178.404	44.318	136.545
4.000	187.793	40.561	124.967
4.200	197.183	37.258	114.791
4.400	206.573	34.328	105.765
4.600	215.962	31.690	97.638
4.800	225.352	29.369	90.486
5.000	234.742	27.304	84.125
5.200	244.131	25.415	78.304
5.325	250.000	24.300	74.869

Table X (61) Axial Compression Load Capacity of Single Equal Angle Section

ISA 130130(8)@ 15.9 kg/m

Section	*Area mm²*	$r_{min} = r_{vv}$ *mm*	*Buckling Class y-y*
ISA 130130(8)@ 15.9 kg/m	*2022.008*	*25.50*	*c*
Eff Length m	*Sl Ratio*	*D.C.S.* f_{cd} *N/mm²*	*F.A.L. kN*
1.000	39.216	199.020	402.419
1.200	47.059	187.412	378.948
1.400	54.902	175.647	355.160
1.600	62.745	163.608	330.817
1.800	70.588	151.059	305.442
2.000	78.431	138.510	280.068
2.200	86.275	126.588	255.963
2.400	94.118	115.235	233.007
2.600	101.961	104.569	211.439
2.800	109.804	94.843	191.774
3.000	117.647	86.265	174.428
3.200	125.490	78.539	158.807
3.400	133.333	71.600	144.776
3.600	141.176	65.376	132.192
3.800	149.020	59.886	121.091
4.000	156.863	55.151	111.516
4.200	164.706	50.853	102.825
4.400	172.549	46.953	94.939
4.600	180.392	43.447	87.850
4.800	188.235	40.388	81.665
5.000	196.078	37.633	76.095
5.200	203.922	35.124	71.020
5.400	211.765	32.824	66.369
5.600	219.608	30.706	62.088
5.800	227.451	28.886	58.408
6.000	235.294	27.188	54.975
6.200	243.137	25.604	51.771
6.375	250.000	24.300	49.135

Table X (62) Axial Compression Load Capacity of Single Equal Angle Section

ISA 130130(10)@ 19.7 kg/m

Section	Area mm^2	$r_{min} = r_{vv}$ mm	Buckling Class y-y
ISA 130130(10)@ 19.7 kg/m	2506.008	25.40	c
Eff Length m	Sl Ratio	D.C.S. f_{cd} N/mm^2	F.A.L. kN
1.000	39.370	198.819	498.242
1.200	47.244	187.134	468.959
1.400	55.118	175.323	439.361
1.600	62.992	163.213	409.012
1.800	70.866	150.614	377.440
2.000	78.740	138.016	345.869
2.200	86.614	126.079	315.954
2.400	94.488	114.717	287.481
2.600	102.362	104.071	260.803
2.800	110.236	94.343	236.423
3.000	118.110	85.760	214.915
3.200	125.984	78.075	195.656
3.400	133.858	71.175	178.365
3.600	141.732	64.987	162.859
3.800	149.606	59.476	149.046
4.000	157.480	54.787	137.296
4.200	165.354	50.516	126.593
4.400	173.228	46.647	116.898
4.600	181.102	43.170	108.185
4.800	188.976	40.099	100.489
5.000	196.850	37.371	93.652
5.200	204.724	34.883	87.416
5.400	212.598	32.598	81.692
5.600	220.472	30.491	76.412
5.800	228.346	28.680	71.873
6.000	236.220	26.994	67.646
6.200	244.094	25.422	63.708
6.350	250.000	24.300	60.896

Table X (63) Axial Compression Load Capacity of Single Equal Angle Section

ISA 130130(12)@ 23.4 kg/m

Section	Area mm^2	$r_{min} = r_{vv}$ mm	Buckling Class y-y
ISA 130130(12)@ 23.4 kg/m	2982.008	25.40	c
Eff Length m	Sl Ratio	D.C.S. f_{cd} N/mm^2	F.A.L. kN
1.000	39.370	198.819	592.880
1.200	47.244	187.134	558.035
1.400	55.118	175.323	522.814
1.600	62.992	163.213	486.701
1.800	70.866	150.614	449.133

Contd...

2.000	78.740	138.016	411.564
2.200	86.614	126.079	375.968
2.400	94.488	114.717	342.086
2.600	102.362	104.071	310.340
2.800	110.236	94.343	281.330
3.000	118.110	85.760	255.737
3.200	125.984	78.075	232.820
3.400	133.858	71.175	212.244
3.600	141.732	64.987	193.793
3.800	149.606	59.476	177.357
4.000	157.480	54.787	163.374
4.200	165.354	50.516	150.638
4.400	173.228	46.647	139.102
4.600	181.102	43.170	128.734
4.800	188.976	40.099	119.576
5.000	196.850	37.371	111.440
5.200	204.724	34.883	104.020
5.400	212.598	32.598	97.209
5.600	220.472	30.491	90.925
5.800	228.346	28.680	85.525
6.000	236.220	26.994	80.495
6.200	244.094	25.422	75.809
6.350	250.000	24.300	72.463

Table X (64) Axial Compression Load Capacity of Single Equal Angle Section

ISA 130130(15)@ 28.9 kg/m

Section	*Area mm²*	$r_{min} = r_{vv}$ *mm*	*Buckling Class y-y*
ISA 130130(15)@ 28.9 kg/m	*3681.008*	*25.30*	*c*
Eff Length m	*Sl Ratio*	*D.C.S.* f_{cd} *N/mm²*	*F.A.L. kN*
1.000	39.526	198.617	731.109
1.200	47.431	186.854	687.810
1.400	55.336	174.996	644.162
1.600	63.241	162.814	599.321
1.800	71.146	150.166	552.762
2.000	79.051	137.518	506.204
2.200	86.957	125.565	462.207
2.400	94.862	114.194	420.348
2.600	102.767	103.569	381.239
2.800	110.672	93.868	345.527
3.000	118.577	85.251	313.810
3.200	126.482	77.607	285.671
3.400	134.387	70.746	260.418
3.600	142.292	64.595	237.776
3.800	150.198	59.083	217.487
4.000	158.103	54.419	200.318

Contd...

4.200	166.008	50.176	184.698
4.400	173.913	46.339	170.575
4.600	181.818	42.891	157.882
4.800	189.723	39.808	146.533
5.000	197.628	37.106	136.589
5.200	205.534	34.640	127.510
5.400	213.439	32.372	119.160
5.600	221.344	30.291	111.501
5.800	229.249	28.473	104.808
6.000	237.154	26.798	98.642
6.200	245.059	25.239	92.904
6.325	250.000	24.300	89.449

able X (65) Axial Compression Load Capacity of Single Equal Angle Section

ISA 150150(10)@ 22.8 kg/m

Section	*Area mm²*	$r_{min} = r_{vv}$ *mm*	*Buckling Class y-y*
ISA 150150(10)@ 22.8 kg/m	*2903.433*	*29.30*	*c*
Eff Length m	*Sl Ratio*	*D.C.S.* f_{cd} *N/mm²*	*F.A.L. kN*
1.000	34.130	205.631	597.037
1.200	40.956	196.567	570.718
1.400	47.782	186.328	540.990
1.600	54.608	176.089	511.262
1.800	61.433	165.706	481.118
2.000	68.259	154.785	449.408
2.200	75.085	143.863	417.698
2.400	81.911	133.133	386.543
2.600	88.737	122.894	356.815
2.800	95.563	113.212	328.702
3.000	102.389	104.038	302.066
3.200	109.215	95.573	277.491
3.400	116.041	88.015	255.547
3.600	122.867	81.005	235.193
3.800	129.693	74.589	216.563
4.000	136.519	69.020	200.394
4.200	143.345	63.859	185.410
4.400	150.171	59.099	171.591
4.600	156.997	55.072	159.898
4.800	163.823	51.312	148.982
5.000	170.648	47.808	138.808
5.200	177.474	44.737	129.890
5.400	184.300	41.923	121.720
5.600	191.126	39.317	114.154
5.800	197.952	36.996	107.416
6.000	204.778	34.867	101.233
6.200	211.604	32.867	95.427

Contd...

6.400	218.430	31.024	90.076
6.600	225.256	29.391	85.335
6.800	232.082	27.863	80.898
7.000	238.908	26.429	76.736
7.200	245.734	25.111	72.907
7.325	250.000	24.300	70.553

Table X (66)* Axial Compression Load Capacity of Single Equal Angle Section

ISA 150150(12)@ 27.2 kg/m

Section	*Area mm²*	$r_{min} = r_{vv}$ *mm*	*Buckling Class y-y*
ISA 150150(12)@ 27.2 kg/m	*3459.433*	*29.30*	*c*
Eff Length m	*Sl Ratio*	*D.C.S.* f_{cd} *N/mm²*	*F.A.L. kN*
1.000	34.130	205.631	711.368
1.200	40.956	196.567	680.009
1.400	47.782	186.328	644.588
1.600	54.608	176.089	609.167
1.800	61.433	165.706	573.251
2.000	68.259	154.785	535.468
2.200	75.085	143.863	497.686
2.400	81.911	133.133	460.565
2.600	88.737	122.894	425.144
2.800	95.563	113.212	391.648
3.000	102.389	104.038	359.911
3.200	109.215	95.573	330.630
3.400	116.041	88.015	304.483
3.600	122.867	81.005	280.232
3.800	129.693	74.589	258.035
4.000	136.519	69.020	238.769
4.200	143.345	63.859	220.915
4.400	150.171	59.099	204.450
4.600	156.997	55.072	190.518
4.800	163.823	51.312	177.511
5.000	170.648	47.808	165.389
5.200	177.474	44.737	154.763
5.400	184.300	41.923	145.029
5.600	191.126	39,317	136.015
5.800	197.952	36.996	127.986
6.000	204.778	34.867	120.619
6.200	211.604	32.867	113.701
6.400	218.430	31.024	107.325
6.600	225.256	29.391	101.677
6.800	232.082	27.863	96.390
7.000	238.908	26.429	91.431
7.200	245.734	25.111	86.868
7.325	250.000	24.300	84.064

Table X (67) Axial Compression Load Capacity of Single Equal Angle Section

ISA 150150(15)@33.6 kg/m

Section	Area mm²	$r_{min} = r_{vv}$ mm	Buckling Class y-y
ISA 150150(15)@33.6 kg/m	4278.433	29.20	c
Eff Length m	Sl Ratio	D.C.S. f_{cd} N/mm²	F.A.L. kN
1.000	34.247	205.479	879.130
1.200	41.096	196.356	840.097
1.400	47.945	186.082	796.140
1.600	54.795	175.808	752.184
1.800	61.644	165.370	707.524
2.000	68.493	154.411	660.637
2.200	75.342	143.452	613.750
2.400	82.192	132.712	567.801
2.600	89.041	122.438	523.844
2.800	95.890	112.753	482.408
3.000	102.740	103.603	443.257
3.200	109.589	95.110	406.920
3.400	116.438	87.582	374.715
3.600	123.288	80.610	344.883
3.800	130.137	74.189	317.413
4.000	136.986	68.641	293.676
4.200	143.836	63.515	271.745
4.400	150.685	58.796	251.554
4.600	157.534	54.755	234.265
4.800	164.384	51.021	218.288
5.000	171.233	47.545	203.419
5.200	178.082	44.463	190.232
5.400	184.932	41.677	178.311
5.600	191.781	39.095	167.263
5.800	198.630	36.766	157.300
6.000	205.479	34.656	148.274
6.200	212.329	32.671	139.782
6.400	219.178	30.822	131.870
6.600	226.027	29.214	124.989
6.800	232.877	27.696	118.495
7.000	239.726	26.258	112.341
7.200	246.575	24.951	106.750
7.300	250.000	24.300	103.966

Table X (68) Axial Compression Load Capacity of Single Equal Angle Section

ISA 150150(18)@ 39.9 kg/m

Section	Area mm^2	$r_{min} = r_{vv}$ mm	Buckling Class y-y
ISA 150150(18)@ 39.9 kg/m	5079.433	29.10	c
Eff Length m	Sl Ratio	D.C.S. f_{cd} N/mm^2	F.A.L. kN
1.000	34.364	205.326	1042.942
1.200	41.237	196.144	996.302
1.400	48.110	185.835	943.937
1.600	54.983	175.526	891.572
1.800	61.856	165.031	838.264
2.000	68.729	154.034	782.407
2.200	75.601	143.038	726.551
2.400	82.474	132.289	671.951
2.600	89.347	121.979	619.586
2.800	96.220	112.292	570.380
3.000	103.093	103.165	524.020
3.200	109.966	94.643	480.731
3.400	116.838	87.146	442.653
3.600	123.711	80.211	407.428
3.800	130.584	73.827	374.998
4.000	137.457	68.260	346.721
4.200	144.330	63.169	320.863
4.400	151.203	58.490	297.098
4.600	158.076	54.435	276.501
4.800	164.948	50.727	257.663
5.000	171.821	47.280	240.158
5.200	178.694	44.188	224.448
5.400	185.567	41.429	210.435
5.600	192.440	38.870	197.440
5.800	199.313	36.534	185.570
6.000	206.186	34.444	174.958
6.200	213.058	32.474	164.951
6.400	219.931	30.619	155.525
6.600	226.804	29.035	147.482
6.800	233.677	27.528	139.826
7.000	240.550	26.096	132.551
7.200	247.423	24.790	125.918
7.275	250.000	24.300	123.430

Table X (69) Axial Compression Load Capacity of Single Equal Angle Section

ISA 200200(12)@ 36.6 kg/m

Section	Area mm^2	$r_{min} = r_{vv}$ mm	Buckling Class y-y
ISA 200200(12)@ 36.6 kg/m	4661.365	39.20	c
Eff Length m	Sl Ratio	D.C.S. f_{cd} N/mm^2	F.A.L. kN
1.000	25.510	216.837	1010.755
1.200	30.612	210.204	979.838
1.400	35.714	203.571	948.921
1.600	40.816	196.776	917.242
1.800	45.918	189.122	881.569
2.000	51.020	181.469	845.895
2.200	56.122	173.816	810.221
2.400	61.224	166.041	773.977
2.600	66.327	157.878	735.925
2.800	71.429	149.714	697.873
3.000	76.531	141.551	659.821
3.200	81.633	133.551	622.530
3.400	86.735	125.898	586.856
3.600	91.837	118.429	552.039
3.800	96.939	111.286	518.743
4.000	102.041	104.469	486.970
4.200	107.143	98.143	457.480
4.400	112.245	92.153	429.559
4.600	117.347	86.592	403.636
4.800	122.449	81.398	379.426
5.000	127.551	76.602	357.070
5.200	132.653	72.151	336.322
5.400	137.755	68.018	317.058
5.600	142.857	64.200	299.260
5.800	147.959	60.629	282.612
6.000	153.061	57.394	267.534
6.200	158.163	54.384	253.502
6.400	163.265	51.602	240.536
6.600	168.367	48.949	228.169
6.800	173.469	46.539	216.934
7.000	178.571	44.243	206.232
7.200	183.673	42.167	196.557
7.400	188.776	40.178	187.282
7.600	193.878	38.382	178.911
7.800	198.980	36.647	170.825
8.000	204.082	35.076	163.500
8.200	209.184	33.545	156.365
8.400	214.286	32.143	149.830
8.600	219.388	30.765	143.408
8.800	224.490	29.567	137.824

Contd...

9.000	229.592	28.394	132.354
9.200	234.694	27.314	127.322
9.400	239.796	26.243	122.328
9.600	244.898	25.269	117.790
9.800	250.000	24.300	113.271

Table X (70) Axial Compression Load Capacity of Single Equal Angle Section

ISA 200200(15)@ 45.4 kg/m

Section	*Area mm²*	$r_{min} = r_{vv}$ *mm*	*Buckling Class y-y*
ISA 200200(15)@ 45.4 kg/m	*5780.365*	*39.10*	*c*
Eff Length m	*Sl Ratio*	*D.C.S.* f_{cd} *N/mm²*	*F.A.L. kN*
1.000	25.575	216.752	1252.905
1.200	30.691	210.102	1214.468
1.400	35.806	203.453	1176.031
1.600	40.921	196.619	1136.529
1.800	46.036	188.946	1092.179
2.000	51.151	181.274	1047.828
2.200	56.266	173.601	1003.477
2.400	61.381	165.790	958.328
2.600	66.496	157.606	911.021
2.800	71.611	149.422	863.714
3.000	76.726	141.238	816.406
3.200	81.841	133.238	770.163
3.400	86.957	125.565	725.813
3.600	92.072	118.100	682.660
3.800	97.187	110.939	641.266
4.000	102.302	104.146	602.001
4.200	107.417	97.803	565.337
4.400	112.532	91.840	530.870
4.600	117.647	86.265	498.641
4.800	122.762	81.104	468.808
5.000	127.877	76.295	441.015
5.200	132.992	71.876	415.471
5.400	138.107	67.733	391.521
5.600	143.223	63.944	369.621
5.800	148.338	60.364	348.924
6.000	153.453	57.163	330.423
6.200	158.568	54.145	312.978
6.400	163.683	51.385	297.024
6.600	168.798	48.725	281.649
6.800	173.913	46.339	267.857
7.000	179.028	44.037	254.552
7.200	184.143	41.984	242.684
7.400	189.258	39.989	231.153
7.600	194.373	38.213	220.885
7.800	199.488	36.474	210.833

Contd...

8.000	204.604	34.919	201.844
8.200	209.719	33.384	192.974
8.400	214.834	31.995	184.942
8.600	219.949	30.614	176.959
8.800	225.064	29.435	170.147
9.000	230.179	28.262	163.367
9.200	235.294	27.188	157.158
9.400	240.409	26.122	150.996
9.600	245.524	25.150	145.378
9.775	250.000	24.300	140.463

able X (71) Axial Compression Load Capacity of Single Equal Angle Section

ISA 200200(18)@ 54.0 kg/m

Section	*Area mm²*	*$r_{min} = r_{vv}$ mm*	*Buckling Class y-y*
ISA 200200(18)@ 54.0 kg/m	*6881.365*	*39.00*	*c*
Eff Length m	*Sl Ratio*	*D.C.S. f_{cd} N/mm²*	*F.A.L. kN*
1.000	25.641	216.667	1490.962
1.200	30.769	210.000	1445.087
1.400	35.897	203.333	1399.211
1.600	41.026	196.462	1351.924
1.800	46.154	188.769	1298.990
2.000	51.282	181.077	1246.056
2.200	56.410	173.385	1193.123
2.400	61.538	165.538	1139.131
2.600	66.667	157.333	1082.668
2.800	71.795	149.128	1026.206
3.000	76.923	140.923	969.743
3.200	82.051	132.923	914.692
3.400	87.179	125.231	861.759
3.600	92.308	117.769	810.413
3.800	97.436	110.590	761.008
4.000	102.564	103.821	714.427
4.200	107.692	97.462	670.668
4.400	112.821	91.526	629.821
4.600	117.949	85.936	591.356
4.800	123.077	80.808	556.067
5.000	128.205	75.987	522.896
5.200	133.333	71.600	492.706
5.400	138.462	67.446	464.122
5.600	143.590	63.687	438.255
5.800	148.718	60.097	413.552
6.000	153.846	56.931	391.761
6.200	158.974	53.905	370.941
6.400	164.103	51.167	352.097
6.600	169.231	48.500	333.746
6.800	174.359	46.138	317.496

Contd...

7.000	179.487	43.831	301.616
7.200	184.615	41.800	287.641
7.400	189.744	39.800	273.878
7.600	194.872	38.044	261.792
7.800	200.000	36.300	249.794
8.000	205.128	34.762	239.207
8.200	210.256	33.231	228.673
8.400	215.385	31.846	219.145
8.600	220.513	30.482	209.758
8.800	225.641	29.303	201.642
9.000	230.769	28.138	193.631
9.200	235.897	27.062	186.220
9.400	241.026	26.005	178.951
9.600	246.154	25.031	172.246
9.750	250.000	24.300	167.217

Table X (72) Axial Compression Load Capacity of Single Equal Angle Section

ISA 200200(25)@ 73.6kg/m

Section	*Area mm²*	*$r_{min} = r_{vv}$ mm*	*Buckling Class y-y*
ISA 200200(25)@ 73.6kg/m	*9380.365*	*38.80*	*c*
Eff Length m	*Sl Ratio*	*D.C.S. f_{cd} N/mm²*	*F.A.L. kN*
1.000	25.773	216.495	2030.801
1.200	30.928	209.794	1967.943
1.400	36.082	203.093	1905.084
1.600	41.237	196.144	1839.905
1.800	46.392	188.412	1767.377
2.000	51.546	180.680	1694.848
2.200	56.701	172.948	1622.320
2.400	61.856	165.031	1548.050
2.600	67.010	156.784	1470.687
2.800	72.165	148.536	1393.323
3.000	77.320	140.289	1315.959
3.200	82.474	132.289	1240.916
3.400	87.629	124.557	1168.387
3.600	92.784	117.103	1098.470
3.800	97.938	109.887	1030.776
4.000	103.093	103.165	967.725
4.200	108.247	96.773	907.768
4.400	113.402	90.892	852.598
4.600	118.557	85.273	799.894
4.800	123.711	80.211	752.412
5.000	128.866	75.366	706.960
5.200	134.021	71.043	666.412
5.400	139.175	66.868	627.247
5.600	144.330	63.169	592.549
5.800	149.485	59.561	558.702

Contd...

6.000	154.639	56.463	529.642
6.200	159.794	53.422	501.115
6.400	164.948	50.727	475.836
6.600	170.103	48.054	450.760
6.800	175.258	45.734	429.002
7.000	180.412	43.439	407.475
7.200	185.567	41.429	388.618
7.400	190.722	39.455	370.099
7.600	195.876	37.702	353.659
7.800	201.031	35.991	337.606
8.000	206.186	34.444	323.100
8.200	211.340	32.938	308.972
8.400	216.495	31.546	295.917
8.600	221.649	30.221	283.480
8.800	226.804	29.035	272.359
9.000	231.959	27.889	261.606
9.200	237.113	26.806	251.452
9.400	242.268	25.769	241.723
9.600	247.423	24.790	232.536
9.700	250.000	25.500	239.199

Factored Axial Compression Load Capacity of Single Unequal Angle Section

able XI (01) Axial Compression Load Capacity of Single Unequal Angle Section

ISA 3020(3)@ 1.1 kg/m

Section	Area mm^2	$r_{min} = r_{vv}$ mm	Buckling Class y-y
ISA 3020(3)@ 1.1 kg/m	141.482	4.10	c
Eff Length m	Sl Ratio	D.C.S. f_{cd} N/mm^2	F.A.L. kN
1.0	243.902	25.459	3.602
1.025	250.000	24.300	3.438

able XI (02) Axial Compression Load Capacity of Single Unequal Angle Section

ISA 3020(4)@ 1.1 kg/m

Section	Area mm^2	$r_{min} = r_{vv}$ mm	Buckling Class y-y
ISA 3020(4)@ 1.1 kg/m	134.482	4.10	c
Eff Length m	Sl Ratio	D.C.S. f_{cd} N/mm^2	F.A.L. kN
1.0	243.902	25.459	4.697
1.025	250.000	24.300	4.483

able XI (03) Axial Compression Load Capacity of Single Unequal Angle Section

ISA 3020(5)@ 1.1 kg/m

Section	Area mm^2	$r_{min} = r_{vv}$ mm	Buckling Class y-y
ISA 3020(5)@ 1.1 kg/m	225.482	4.10	c
Eff Length m	Sl Ratio	D.C.S. f_{cd} N/mm^2	F.A.L. kN
1.0	243.902	25.459	5.740
1.025	250.000	24.300	5.479

able XI (04) Axial Compression Load Capacity of Single Unequal Angle Section

ISA 4025(3)@ 1.5 kg/m

Section	Area mm^2	$r_{min} = r_{vv}$ mm	Buckling Class y-y
ISA 4025(3)@ 1.5 kg/m	187.502	5.20	c
Eff Length m	Sl Ratio	D.C.S. f_{cd} N/mm^2	F.A.L. kN
1.0	192.308	38.915	7.297
1.1	211.538	32.885	6.166
1.2	230.769	28.138	5.276
1.300	250.000	24.300	4.556

able XI (05) Axial Compression Load Capacity of Single Unequal Angle Section

ISA 4025(4)@ 1.9 kg/m

Table XI	Area mm^2	$r_{min} = r_{vv}$ mm	Buckling Class y-y
ISA 4025(4)@ 1.9 kg/m	245.502	5.20	c
Eff Length m	Sl Ratio	D.C.S. f_{cd} N/mm^2	F.A.L. kN
1.0	192.308	38.915	9.554
1.1	211.538	32.885	8.073
1.2	230.769	28.138	6.908
1.300	250.000	24.300	5.966

Table XI (06) Axial Compression Load Capacity of Single Unequal Angle Section

ISA 4025(5)@ 2.4 kg/m

Section	Area mm^2	$r_{min} = r_{vv}$ mm	Buckling Class y-y
ISA 4025(5)@ 2.4 kg/m	301.502	5.20	c
Eff Length m	Sl Ratio	D.C.S. f_{cd} N/mm²	F.A.L. kN
1.0	192.308	38.915	11.733
1.1	211.538	32.885	9.915
1.2	230.769	28.138	8.484
1.300	250.000	24.300	7.327

Table XI (07) Axial Compression Load Capacity of Single Unequal Angle Section

ISA 4025(6)@ 2.8 kg/m

Section	Area mm^2	$r_{min} = r_{vv}$ mm	Buckling Class y-y
ISA 4025(6)@ 2.8 kg/m	355.502	5.20	c
Eff Length m	Sl Ratio	D.C.S. f_{cd} N/mm²	F.A.L. kN
1.0	192.308	38.915	13.835
1.1	211.538	32.885	11.691
1.2	230.769	28.138	10.003
1.300	250.000	24.300	8.639

Table XI (08) Axial Compression Load Capacity of Single Unequal Angle Section

ISA 4530(3)@ 1.7 kg/m

Section	Area mm^2	$r_{min} = r_{vv}$ mm	Buckling Class y-y
ISA 4530(3)@ 1.7 kg/m	217.502	6.30	c
Eff Length m	Sl Ratio	D.C.S. f_{cd} N/mm²	F.A.L. kN
1.0	158.730	54.049	11.756
1.1	174.603	46.029	10.011
1.2	190.476	39.538	8.600
1.3	206.349	34.395	7.481
1.4	222.222	30.089	6.544
1.5	238.095	26.600	5.786
1.575	250.000	24.300	5.285

Table XI (09) Axial Compression Load Capacity of Single Unequal Angle Section

ISA 4530(4)@ 2.2 kg/m

Section	Area mm^2	$r_{min} = r_{vv}$ mm	Buckling Class y-y
ISA 4530(4)@ 2.2 kg/m	285.502	6.30	c
Eff Length m	Sl Ratio	D.C.S. f_{cd} N/mm²	F.A.L. kN
1.0	158.730	54.049	15.431
1.1	174.603	46.029	13.141
1.2	190.476	39.538	11.288
1.3	206.349	34.395	9.820
1.4	222.222	30.089	8.590
1.5	238.095	26.600	7.594
1.575	250.000	24.300	6.938

able XI (10) Axial Compression Load Capacity Of Single Unequal Angle Section

ISA 4530(5)@ 2.8 kg/m

Section	Area mm²	$r_{min} = r_{vv}$ mm	Buckling Class y-y
ISA 4530(5)@ 2.8 kg/m	351.502	6.30	c
Eff Length m	Sl Ratio	D.C.S. f_{cd} N/mm²	F.A.L. kN
1.0	158.730	54.049	18.998
1.1	174.603	46.029	16.179
1.2	190.476	39.538	13.898
1.3	206.349	34.395	12.090
1.4	222.222	30.089	10.576
1.5	238.095	26.600	9.350
1.575	250.000	24.300	8.542

able XI (11) Axial Compression Load Capacity of Single Unequal Angle Section

ISA 4530(6)@ 3.3 kg/m

Section	Area mm²	$r_{min} = r_{vv}$ mm	Buckling Class y-y
ISA 4530(6)@ 3.3 kg/m	415.502	6.30	c
Eff Length m	Sl Ratio	D.C.S. f_{cd} N/mm²	F.A.L. kN
1.0	158.730	54.049	22.458
1.1	174.603	46.029	19.125
1.2	190.476	39.538	16.428
1.3	206.349	34.395	14.291
1.4	222.222	30.089	12.502
1.5	238.095	26.600	11.052
1.575	250.000	24.300	10.097

able XI (12) Axial Compression Load Capacity of Single Unequal Angle Section

ISA 5030(3)@ 1.8 kg/m

Section	Area mm²	$r_{min} = r_{vv}$ mm	Buckling Class y-y
ISA 5030(3)@ 1.8 kg/m	233.629	6.50	c
Eff Length m	Sl Ratio	D.C.S. f_{cd} N/mm²	F.A.L. kN
1.0	153.846	56.931	13.301
1.1	169.231	48.500	11.331
1.2	184.615	41.800	9.766
1.3	200.000	36.300	8.481
1.4	215.385	31.846	7.440
1.5	230.769	28.138	6.574
1.6	246.154	25.031	5.848
1.625	250.000	24.300	5.677

Table XI (13) Axial Compression Load Capacity of Single Unequal Angle Section

ISA 5030(4)@ 2.4 kg/m

Section	Area mm^2	$r_{min} = r_{vv}$ mm	Buckling Class y-y
ISA 5030(4)@ 2.4 kg/m	306.629	6.30	c
Eff Length m	Sl Ratio	D.C.S. f_{cd} N/mm^2	F.A.L. kN
1.0	158.730	54.049	16.573
1.1	174.603	46.029	14.114
1.2	190.476	39.538	12.124
1.3	206.349	34.395	10.547
1.4	222.222	29.633	9.086
1.5	238.095	26.600	8.156
1.575	250.000	24.300	7.451

Table XI (14) Axial Compression Load Capacity of Single Unequal Angle Section

ISA 5030(5)@ 3.0 kg/m

Section	Area mm^2	$r_{min} = r_{vv}$ mm	Buckling Class y-y
ISA 5030(5)@ 3.0 kg/m	377.629	6.30	c
Eff Length m	Sl Ratio	D.C.S. f_{cd} N/mm^2	F.A.L. kN
1.0	158.730	54.049	20.411
1.1	174.603	46.029	17.382
1.2	190.476	39.538	14.931
1.3	206.349	34.395	12.989
1.4	222.222	29.633	11.190
1.5	238.095	26.600	10.045
1.575	250.000	24.300	9.176

Table XI (15) Axial Compression Load Capacity of Single Unequal Angle Section

ISA 5030(6)@ 3.5 kg/m

Table XI	Area mm^2	$r_{min} = r_{vv}$ mm	Buckling Class y-y
ISA 5030(6)@ 3.5 kg/m	446.629	6.30	c
Eff Length m	Sl Ratio	D.C.S. f_{cd} N/mm^2	F.A.L. kN
1.0	158.730	54.049	24.140
1.1	174.603	46.029	20.558
1.2	190.476	39.538	17.659
1.3	206.349	34.395	15.362
1.4	222.222	29.633	13.235
1.5	238.095	26.600	11.880
1.575	250.000	24.300	10.853

Table XI (16) Axial Compression Load Capacity of Single Unequal Angle Section

ISA 6040(5)@ 3.7 kg/m

Section	Area mm²	$r_{min} = r_{vv}$ mm	Buckling Class y-y
ISA 6040(5)@ 3.7 kg/m	475.858	8.50	c
Eff Length m	Sl Ratio	D.C.S. f_{cd} N/mm²	F.A.L. kN
1.0	117.647	86.265	41.050
1.1	129.412	74.853	35.619
1.2	141.176	65.376	31.110
1.3	152.941	57.465	27.345
1.4	164.706	50.853	24.199
1.5	176.471	45.188	21.503
1.6	188.235	40.388	19.219
1.7	200.000	36.300	17.274
1.8	211.765	32.824	15.619
1.9	223.529	29.788	14.175
2.0	235.294	27.188	12.938
2.1	247.059	24.859	11.829
2.125	250.000	24.300	11.563

Table XI (17) Axial Compression Load Capacity of Single Unequal Angle Section

ISA 6040(6)@ 4.4 kg/m

Section	Area mm²	$r_{min} = r_{vv}$ mm	Buckling Class y-y
ISA 6040(6)@ 4.4 kg/m	564.858	8.50	c
Eff Length m	Sl Ratio	D.C.S. f_{cd} N/mm²	F.A.L. kN
1.0	117.647	86.265	48.727
1.1	129.412	74.853	42.281
1.2	141.176	65.376	36.928
1.3	152.941	57.465	32.459
1.4	164.706	50.853	28.725
1.5	176.471	45.188	25.525
1.6	188.235	40.388	22.814
1.7	200.000	36.300	20.504
1.8	211.765	32.824	18.541
1.9	223.529	29.788	16.826
2.0	235.294	27.188	15.358
2.1	247.059	24.859	14.042
2.125	250.000	24.300	13.726

Table XI (18) Axial Compression Load Capacity of Single Unequal Angle Section

ISA 6040(8)@ 5.8 kg/m

Section	Area mm²	$r_{min} = r_{vv}$ mm	Buckling Class y-y
ISA 6040(8)@ 5.8 kg/m	736.858	8.40	c
Eff Length m	Sl Ratio	D.C.S. f_{cd} N/mm²	F.A.L. kN
1.0	119.048	84.738	62.440
1.1	130.952	73.529	54.180

Contd...

1.2	142.857	64.200	47.306
1.3	154.762	56.390	41.552
1.4	166.667	49.833	36.720
1.5	178.571	48.843	35.990
1.6	190.476	39.538	29.134
1.7	202.381	35.586	26.222
1.8	214.286	32.143	23.685
1.9	226.190	29.176	21.499
2.0	238.095	26.600	19.600
2.100	250.000	24.300	17.906

Table XI (19) Axial Compression Load Capacity of Single Unequal Angle Section

ISA 6545(5)@ 4.1 kg/m

Section	*Area mm^2*	*$r_{min} = r_{vv}$ mm*	*Buckling Class y-y*
ISA 6545(5)@ 4.1 kg/m	*525.858*	*9.60*	*c*
Eff Length m	*Sl Ratio*	*D.C.S. f_{cd} N/mm^2*	*F.A.L. kN*
1.0	104.167	101.833	53.550
1.1	114.583	89.604	47.119
1.2	125.000	79.000	41.543
1.3	135.417	69.913	36.764
1.4	145.833	62.117	32.665
1.5	156.250	55.513	29.192
1.6	166.667	49.833	26.205
1.7	177.083	44.913	23.618
1.8	187.500	40.675	21.389
1.9	197.917	37.008	19.461
2.0	208.333	33.800	17.774
2.1	218.750	30.938	16.269
2.2	229.167	28.492	14.983
2.3	239.583	26.288	13.824
2.400	250.000	24.300	12.778

Table XI (20) Axial Compression Load Capacity Of Single Unequal Angle Section

ISA 6545(6)@ 4.9 kg/m

Section	*Area mm^2*	*$r_{min} = r_{vv}$ mm*	*Buckling Class y-y*
ISA 6545(6)@ 4.9 kg/m	*624.858*	*9.50*	*c*
Eff Length m	*Sl Ratio*	*D.C.S. f_{cd} N/mm^2*	*F.A.L. kN*
1.0	105.263	100.474	62.782
1.1	115.789	88.289	55.168
1.2	126.316	77.763	48.591
1.3	136.842	68.758	42.964
1.4	147.368	61.042	38.143
1.5	157.895	54.542	34.081
1.6	168.421	48.921	30.569
1.7	178.947	44.074	27.540

Contd...

1.8	189.474	39.905	24.935
1.9	200.000	36.300	22.682
2.0	210.526	33.158	20.719
2.1	221.053	30.358	18.969
2.2	231.579	27.968	17.476
2.3	242.105	25.800	16.121
2.375	250.000	24.300	15.184

able XI (21) Axial Compression Load Capacity of Single Unequal Angle Section

ISA 6545(8)@ 6.4 kg/m

Section	*Area mm²*	*$r_{min} = r_{vv}$ mm*	*Buckling Class y-y*
ISA 6545(8)@ 6.4 kg/m	*816.858*	*9.50*	*c*
Eff Length m	*Sl Ratio*	*D.C.S. f_{cd} N/mm²*	*F.A.L. kN*
1.0	105.263	100.474	82.073
1.1	115.789	88.289	72.120
1.2	126.316	77.763	63.521
1.3	136.842	68.758	56.165
1.4	147.368	61.042	49.863
1.5	157.895	54.542	44.553
1.6	168.421	48.921	39.962
1.7	178.947	44.074	36.002
1.8	189.474	39.905	32.597
1.9	200.000	36.300	29.652
2.0	210.526	33.158	27.085
2.1	221.053	30.358	24.798
2.2	231.579	27.968	22.846
2.3	242.105	25.800	21.075
2.375	250.000	24.300	19.850

able XI (22) Axial Compression Load Capacity of Single Unequal Angle Section

ISA 7045(5)@ 4.3 kg/m

Section	*Area mm²*	*$r_{min} = r_{vv}$ mm*	*Buckling Class y-y*
ISA 7045(5)@ 4.3 kg/m	*552.200*	*9.60*	*c*
Eff Length m	*Sl Ratio*	*D.C.S. f_{cd} N/mm²*	*F.A.L. kN*
1.0	104.167	101.833	56.232
1.1	114.583	89.604	49.479
1.2	125.000	79.000	43.624
1.3	135.417	69.913	38.606
1.4	145.833	62.117	34.301
1.5	156.250	55.513	30.654
1.6	166.667	49.833	27.518
1.7	177.083	44.913	24.801
1.8	187.500	40.675	22.461
1.9	197.917	37.008	20.436
2.0	208.333	33.800	18.664
2.1	218.750	30.938	17.084

Contd...

2.2	229.167	28.492	15.733
2.3	239.583	26.288	14.516
2.400	250.000	24.300	13.418

Table XI (23) Axial Compression Load Capacity of Single Unequal Angle Section

ISA 7045(6)@ 5.2 kg/m

Section	*Area mm²*	$r_{min} = r_{vv}$ *mm*	*Buckling Class y-y*
ISA 7045(6)@ 5.2 kg/m	*656.200*	*9.60*	*c*
Eff Length m	*Sl Ratio*	*D.C.S.* f_{cd} *N/mm²*	*F.A.L. kN*
1.0	104.167	101.833	66.823
1.1	114.583	89.604	58.798
1.2	125.000	79.000	51.840
1.3	135.417	69.913	45.877
1.4	145.833	62.117	40.761
1.5	156.250	55.513	36.427
1.6	166.667	49.833	32.701
1.7	177.083	44.913	29.472
1.8	187.500	40.675	26.691
1.9	197.917	37.008	24.285
2.0	208.333	33.800	22.180
2.1	218.750	30.938	20.301
2.2	229.167	28.492	18.696
2.3	239.583	26.288	17.250
2.400	250.000	24.300	15.946

Table XI (24) Axial Compression Load Capacity of Single Unequal Angle Section

ISA 7045(8)@ 6.7 kg/m

Section	*Area mm²*	$r_{min} = r_{vv}$ *mm*	*Buckling Class y-y*
ISA 7045(8)@ 6.7 kg/m	*858.200*	*9.50*	*c*
Eff Length m	*Sl Ratio*	*D.C.S.* f_{cd} *N/mm²*	*F.A.L. kN*
1.0	105.263	100.474	86.226
1.1	115.789	88.289	75.770
1.2	126.316	77.763	66.736
1.3	136.842	68.758	59.008
1.4	147.368	61.042	52.386
1.5	157.895	54.542	46.808
1.6	168.421	48.921	41.984
1.7	178.947	44.074	37.824
1.8	189.474	39.905	34.247
1.9	200.000	36.300	31.153
2.0	210.526	33.158	28.456
2.1	221.053	30.358	26.053
2.2	231.579	27.968	24.002
2.3	242.105	25.800	22.142
2.375	250.000	24.300	20.854

Table XI (25) Axial Compression Load Capacity of Single Unequal Angle Section

ISA 7045(10)@ 8.3 kg/m

Section	Area mm^2	$r_{min} = r_{vv}$ mm	Buckling Class y-y
ISA 7045(10)@ 8.3 kg/m	1052.200	9.50	c
Eff Length m	Sl Ratio	D.C.S. f_{cd} N/mm^2	F.A.L. kN
1.0	105.263	100.474	105.718
1.1	115.789	88.289	92.898
1.2	126.316	77.763	81.822
1.3	136.842	68.758	72.347
1.4	147.368	61.042	64.228
1.5	157.895	54.542	57.389
1.6	168.421	48.921	51.475
1.7	178.947	44.074	46.374
1.8	189.474	39.905	41.988
1.9	200.000	36.300	38.195
2.0	210.526	33.158	34.889
2.1	221.053	30.358	31.943
2.2	231.579	27.968	29.428
2.3	242.105	25.800	27.147
2.4	250.000	24.300	25.568

Table XI (26) Axial Compression Load Capacity of Single Unequal Angle Section

ISA 7550(5)@ 4.7 kg/m

Section	Area mm^2	$r_{min} = r_{vv}$ mm	Buckling Class y-y
ISA 7550(5)@ 4.7 kg/m	602.200	10.70	c
Eff Length m	Sl Ratio	D.C.S. f_{cd} N/mm^2	F.A.L. kN
1.0	93.458	116.159	69.951
1.1	102.804	103.523	62.342
1.2	112.150	92.257	55.557
1.3	121.495	82.294	49.558
1.4	130.841	73.619	44.333
1.5	140.187	66.069	39.787
1.6	149.533	59.527	35.847
1.7	158.879	53.962	32.496
1.8	168.224	49.023	29.522
1.9	177.570	44.693	26.914
2.0	186.916	40.903	24.632
2.1	196.262	37.571	22.625
2.2	205.607	34.618	20.847
2.3	214.953	31.963	19.248
2.4	224.299	29.611	17.832
2.5	233.645	27.535	16.581
2.6	242.991	25.632	15.435
2.675	250.000	24.300	14.633

Table XI (27) Axial Compression Load Capacity of Single Unequal Angle Section

ISA 7550(6)@ 5.6 kg/m

Section	Area mm²	$r_{min} = r_{vv}$ mm	Buckling Class y-y
ISA 7550(6)@ 5.6 kg/m	716.200	10.70	c
Eff Length m	Sl Ratio	D.C.S. f_{cd} N/mm²	F.A.L. kN
1.0	93.458	116.159	83.193
1.1	102.804	103.523	74.143
1.2	112.150	92.257	66.074
1.3	121.495	82.294	58.939
1.4	130.841	73.619	52.726
1.5	140.187	66.069	47.319
1.6	149.533	59.527	42.633
1.7	158.879	53.962	38.647
1.8	168.224	49.023	35.111
1.9	177.570	44.693	32.009
2.0	186.916	40.903	29.295
2.1	196.262	37.571	26.908
2.2	205.607	34.618	24.793
2.3	214.953	31.963	22.892
2.4	224.299	29.611	21.208
2.5	233.645	27.535	19.720
2.6	242.991	25.632	18.357
2.675	250.000	24.300	17.404

Table XI (28) Axial Compression Load Capacity of Single Unequal Angle Section

ISA 7550(8)@ 7.4 kg/m

Section	Area mm²	$r_{min} = r_{vv}$ mm	Buckling Class y-y
ISA 7550(8)@ 7.4 kg/m	938.200	10.60	c
Eff Length m	Sl Ratio	D.C.S. f_{cd} N/mm²	F.A.L. kN
1.0	94.340	114.925	107.822
1.1	103.774	102.321	95.997
1.2	113.208	91.104	85.474
1.3	122.642	81.217	76.198
1.4	132.075	72.619	68.131
1.5	141.509	65.143	61.118
1.6	150.943	58.643	55.019
1.7	160.377	53.104	49.822
1.8	169.811	48.198	45.219
1.9	179.245	43.940	41.224
2.0	188.679	40.215	37.730
2.1	198.113	36.942	34.659
2.2	207.547	34.036	31.932
2.3	216.981	31.415	29.474
2.4	226.415	29.125	27.325

Contd...

2.5	235.849	27.072	25.399
2.6	245.283	25.196	23.639
2.650	250.000	24.300	22.798

able XI (29) Axial Compression Load Capacity of Single Unequal Angle Section

ISA 7550(10)@ 9.0 kg/m

Section	*Area mm²*	$r_{min} = r_{vv}$ *mm*	*Buckling Class y-y*
ISA 7550(10)@ 9.0 kg/m	*1152.200*	*10.60*	*c*
Eff Length m	*Sl Ratio*	*D.C.S.* f_{cd} *N/mm²*	*F.A.L. kN*
1.0	94.340	114.925	132.416
1.1	103.774	102.321	117.894
1.2	113.208	91.104	104.970
1.3	122.642	81.217	93.578
1.4	132.075	72.619	83.671
1.5	141.509	65.143	75.058
1.6	150.943	58.643	67.569
1.7	160.377	53.104	61.186
1.8	169.811	48.198	55.534
1.9	179.245	43.940	50.627
2.0	188.679	40.215	46.336
2.1	198.113	36.942	42.564
2.2	207.547	34.036	39.216
2.3	216.981	31.415	36.196
2.4	226.415	29.125	33.557
2.5	235.849	27.072	31.192
2.6	245.283	25.196	29.031
2.650	250.000	24.300	27.998

ble XI (30) Axial Compression Load Capacity of Single Unequal Angle Section

ISA 8050(5)@ 4.9 kg/m

Section	*Area mm²*	$r_{min} = r_{vv}$ *mm*	*Buckling Class y-y*
ISA 8050(5)@ 4.9 kg/m	*626.824*	*10.70*	*c*
Eff Length m	*Sl Ratio*	*D.C.S.* f_{cd} *N/mm²*	*F.A.L. kN*
1.0	93.458	116.159	72.811
1.1	102.804	103.523	64.891
1.2	112.150	92.257	57.829
1.3	121.495	82.294	51.584
1.4	130.841	73.619	46.146
1.5	140.187	66.069	41.414
1.6	149.533	59.527	37.313
1.7	158.879	53.962	33.824
1.8	168.224	49.023	30.729
1.9	177.570	44.693	28.015
2.0	186.916	40.903	25.639

Contd...

2.1	196.262	37.571	23.550
2.2	205.607	34.618	21.699
2.3	214.953	31.963	20.035
2.4	224.299	29.611	18.561
2.5	233.645	27.535	17.259
2.6	242.991	25.632	16.067
2.675	250.000	24.300	15.232

Table XI (31) Axial Compression Load Capacity of Single Unequal Angle Section

ISA 8050(6)@ 5.9 kg/m

Section	*Area mm²*	*$r_{min} = r_{vv}$ mm*	*Buckling Class y-y*
ISA 8050(6)@ 5.9 kg/m	*716.200*	*10.70*	*c*
Eff Length m	*Sl Ratio*	*D.C.S. f_{cd} N/mm²*	*F.A.L. kN*
1.0	93.458	116.159	83.193
1.1	102.804	103.523	74.143
1.2	112.150	92.257	66.074
1.3	121.495	82.294	58.939
1.4	130.841	73.619	52.726
1.5	140.187	66.069	47.319
1.6	149.533	59.527	42.633
1.7	158.879	53.962	38.647
1.8	168.224	49.023	35.111
1.9	177.570	44.693	32.009
2.0	186.916	40.903	29.295
2.1	196.262	37.571	26.908
2.2	205.607	34.618	24.793
2.3	214.953	31.963	22.892
2.4	224.299	29.611	21.208
2.5	233.645	27.535	19.720
2.6	242.991	25.632	18.357
2.675	250.000	24.300	17.404

Table XI (32) Axial Compression Load Capacity of Single Unequal Angle Section

ISA 8050(8)@ 7.7 kg/m

Section	*Area mm²*	*$r_{min} = r_{vv}$ mm*	*Buckling Class y-y*
ISA 8050(8)@ 7.7 kg/m	*977.824*	*10.60*	*c*
Eff Length m	*Sl Ratio*	*D.C.S. f_{cd} N/mm²*	*F.A.L. kN*
1.0	94.340	114.925	112.376
1.1	103.774	102.321	100.052
1.2	113.208	91.104	89.083
1.3	122.642	81.217	79.416
1.4	132.075	72.619	71.008
1.5	141.509	65.143	63.699
1.6	150.943	58.643	57.343
1.7	160.377	53.104	51.926

Contd...

1.8	169.811	48.198	47.129
1.9	179.245	43.940	42.965
2.0	188.679	40.215	39.323
2.1	198.113	36.942	36.122
2.2	207.547	34.036	33.281
2.3	216.981	31.415	30.718
2.4	226.415	29.125	28.479
2.5	235.849	27.072	26.471
2.6	245.283	25.196	24.637
2.650	250.000	24.300	23.761

able XI (33) Axial Compression Load Capacity of Single Unequal Angle Section

ISA 8050(10)@ 9.4 kg/m

Section	*Area mm²*	*$r_{min} = r_{vv}$ mm*	*Buckling Class y-y*
ISA 8050(10)@ 9.4 kg/m	*1201.824*	*10.60*	*c*
Eff Length m	*Sl Ratio*	*D.C.S. f_{cd} N/mm²*	*F.A.L. kN*
1.0	94.340	114.925	138.119
1.1	103.774	102.321	122.972
1.2	113.208	91.104	109.491
1.3	122.642	81.217	97.609
1.4	132.075	72.619	87.275
1.5	141.509	65.143	78.291
1.6	150.943	58.643	70.479
1.7	160.377	53.104	63.821
1.8	169.811	48.198	57.926
1.9	179.245	43.940	52.808
2.0	188.679	40.215	48.331
2.1	198.113	36.942	44.397
2.2	207.547	34.036	40.905
2.3	216.981	31.415	37.755
2.4	226.415	29.125	35.003
2.5	235.849	27.072	32.535
2.6	245.283	25.196	30.281
2.650	250.000	24.300	29.204

ble XI (34) Axial Compression Load Capacity of Single Unequal Angle Section

ISA 9060(6)@ 6.8 kg/m

Section	*Area mm²*	*$r_{min} = r_{vv}$ mm*	*Buckling Class y-y*
ISA 9060(6)@ 6.8 kg/m	*865.341*	*12.80*	*c*
Eff Length m	*Sl Ratio*	*D.C.S. f_{cd} N/mm²*	*F.A.L. kN*
1.0	78.125	139.000	120.282
1.1	85.938	127.094	109.979
1.2	93.750	115.750	100.163
1.3	101.563	105.063	90.915
1.4	109.375	95.375	82.532

Contd...

1.5	117.188	86.766	75.082
1.6	125.000	79.000	68.362
1.7	132.813	72.022	62.324
1.8	140.625	65.763	56.907
1.9	148.438	60.294	52.175
2.0	156.250	55.513	48.037
2.1	164.063	51.188	44.295
2.2	171.875	47.256	40.893
2.3	179.688	43.741	37.851
2.4	187.500	40.675	35.198
2.5	195.313	37.894	32.791
2.6	203.125	35.363	30.601
2.7	210.938	33.047	28.597
2.8	218.750	30.938	26.771
2.9	226.563	29.091	25.173
3.0	234.375	27.381	23.694
3.1	242.188	25.784	22.312
3.200	250.000	24.300	21.028

Table XI (35) Axial Compression Load Capacity of Single Unequal Angle Section

ISA 9060(8)@ 8.9 kg/m

Section	*Area mm²*	$r_{min} = r_{vv}$ *mm*	*Buckling Class y-y*
ISA 9060(8)@ 8.9 kg/m	*1137.341*	*12.80*	*c*
Eff Length m	*Sl Ratio*	*D.C.S.* f_{cd} *N/mm²*	*F.A.L. kN*
1.0	78.125	139.000	158.090
1.1	85.938	127.094	144.549
1.2	93.750	115.750	131.647
1.3	101.563	105.063	119.492
1.4	109.375	95.375	108.474
1.5	117.188	86.766	98.682
1.6	125.000	79.000	89.850
1.7	132.813	72.022	81.913
1.8	140.625	65.763	74.794
1.9	148.438	60.294	68.575
2.0	156.250	55.513	63.137
2.1	164.063	51.188	58.218
2.2	171.875	47.256	53.746
2.3	179.688	43.741	49.748
2.4	187.500	40.675	46.261
2.5	195.313	37.894	43.098
2.6	203.125	35.363	40.219
2.7	210.938	33.047	37.586
2.8	218.750	30.938	35.186
2.9	226.563	29.091	33.086

Contd...

3.0	234.375	27.381	31.142
3.1	242.188	25.784	29.326
3.200	250.000	24.300	27.637

able XI (36) Axial Compression Load Capacity of Single Unequal Angle Section

ISA 9060(10)@ 11.0 kg/m

Section	*Area mm²*	$r_{min} = r_{vv}$ *mm*	*Buckling Class y-y*
ISA 9060(10)@ 11.0 kg/m	*1401.341*	*12.70*	*c*
Eff Length m	*Sl Ratio*	*D.C.S.* f_{cd} *N/mm²*	*F.A.L. kN*
1.0	78.740	138.016	193.407
1.1	86.614	126.079	176.679
1.2	94.488	114.717	160.757
1.3	102.362	104.071	145.839
1.4	110.236	94.343	132.206
1.5	118.110	85.760	120.179
1.6	125.984	78.075	109.409
1.7	133.858	71.175	99.740
1.8	141.732	64.987	91.070
1.9	149.606	59.476	83.346
2.0	157.480	54.787	76.775
2.1	165.354	50.516	70.790
2.2	173.228	46.647	65.369
2.3	181.102	43.170	60.496
2.4	188.976	40.099	56.193
2.5	196.850	37.371	52.369
2.6	204.724	34.883	48.883
2.7	212.598	32.598	45.682
2.8	220.472	30.491	42.729
2.9	228.346	28.680	40.191
3.0	236.220	26.994	37.827
3.1	244.094	25.422	35.625
3.175	250.000	24.300	34.053

ble XI (37) Axial Compression Load Capacity of Single Unequal Angle Section

ISA 9060(12)@ 13.0 kg/m

Section	*Area mm²*	$r_{min} = r_{vv}$ *mm*	*Buckling Class y-y*
ISA 9060(12)@ 13.0 kg/m	*1657.341*	*12.70*	*c*
Eff Length m	*Sl Ratio*	*D.C.S.* f_{cd} *N/mm²*	*F.A.L. kN*
1.0	78.740	138.016	228.739
1.1	86.614	126.079	208.955
1.2	94.488	114.717	190.124
1.3	102.362	104.071	172.481
1.4	110.236	94.343	156.358
1.5	118.110	85.760	142.133

Contd...

1.6	125.984	78.075	129.397
1.7	133.858	71.175	117.961
1.8	141.732	64.987	107.706
1.9	149.606	59.476	98.571
2.0	157.480	54.787	90.800
2.1	165.354	50.516	83.722
2.2	173.228	46.647	77.310
2.3	181.102	43.170	71.548
2.4	188.976	40.099	66.458
2.5	196.850	37.371	61.936
2.6	204.724	34.883	57.813
2.7	212.598	32.598	54.027
2.8	220.472	30.491	50.535
2.9	228.346	28.680	47.533
3.0	236.220	26.994	44.738
3.1	244.094	25.422	42.133
3.175	250.000	24.300	40.273

Table XI (38) Axial Compression Load Capacity of Single Unequal Angle Section

ISA 10065(6)@ 7.5 kg/m

Section	*Area mm²*	$r_{min} = r_{vv}$ *mm*	*Buckling Class y-y*
ISA 10065(6)@ 7.5 kg/m	*954.751*	*13.90*	*c*
Eff Length m	*Sl Ratio*	*D.C.S.* f_{cd} *N/mm²*	*F.A.L. kN*
1.0	71.942	148.892	142.155
1.1	79.137	137.381	131.165
1.2	86.331	126.504	120.779
1.3	93.525	116.065	110.813
1.4	100.719	106.108	101.307
1.5	107.914	97.187	92.789
1.6	115.108	89.032	85.004
1.7	122.302	81.536	77.847
1.8	129.496	74.773	71.390
1.9	136.691	68.881	65.764
2.0	143.885	63.481	60.608
2.1	151.079	58.563	55.913
2.2	158.273	54.319	51.861
2.3	165.468	50.457	48.174
2.4	172.662	46.902	44.780
2.5	179.856	43.665	41.689
2.6	187.050	40.850	39.002
2.7	194.245	38.257	36.526
2.8	201.439	35.868	34.245
2.9	208.633	33.710	32.185
3.0	215.827	31.727	30.291
3.1	223.022	29.905	28.552

Contd...

3.2	230.216	28.255	26.976
3.3	237.410	26.744	25.534
3.4	244.604	25.325	24.179
3.475	250.000	24.300	23.200

able XI (39) Axial Compression Load Capacity of Single Unequal Angle Section

ISA 10065(8)@ 9.9 kg/m

Section	Area mm²	$r_{min} = r_{vv}$ mm	Buckling Class y-y
ISA 10065(8)@ 9.9 kg/m	1256.751	13.90	c
Eff Length m	Sl Ratio	D.C.S. f_{cd} N/mm²	F.A.L. kN
1.0	71.942	148.892	187.120
1.1	79.137	137.381	172.654
1.2	86.331	126.504	158.984
1.3	93.525	116.065	145.865
1.4	100.719	106.108	133.351
1.5	107.914	97.187	122.140
1.6	115.108	89.032	111.892
1.7	122.302	81.536	102.470
1.8	129.496	74.773	93.972
1.9	136.691	68.881	86.566
2.0	143.885	63.481	79.779
2.1	151.079	58.563	73.600
2.2	158.273	54.319	68.265
2.3	165.468	50.457	63.412
2.4	172.662	46.902	58.944
2.5	179.856	43.665	54.876
2.6	187.050	40.850	51.339
2.7	194.245	38.257	48.079
2.8	201.439	35.868	45.078
2.9	208.633	33.710	42.365
3.000	215.827	31.727	39.872
3.1	223.022	29.905	37.583
3.2	230.216	28.255	35.509
3.3	237.410	26.744	33.610
3.4	244.604	25.325	31.827
3.475	250.000	24.300	30.539

ble XI (40) Axial Compression Load Capacity of Single Unequal Angle Section

ISA 10065(10)@ 12.2 kg/m

Section	Area mm²	$r_{min} = r_{vv}$ mm	Buckling Class y-y
ISA 10065(10)@ 12.2 kg/m	1550.751	13.80	c
Eff Length m	Sl Ratio	D.C.S. f_{cd} N/mm²	F.A.L. kN
1.0	72.464	148.058	229.601
1.1	79.710	136.464	211.621
1.2	86.957	125.565	194.720

Contd...

1.3	94.203	115.116	178.516
1.4	101.449	105.203	163.144
1.5	108.696	96.217	149.209
1.6	115.942	88.123	136.657
1.7	123.188	80.703	125.150
1.8	130.435	73.948	114.675
1.9	137.681	68.078	105.572
2.0	144.928	62.751	97.311
2.1	152.174	57.917	89.815
2.2	159.420	53.642	83.185
2.3	166.667	49.833	77.279
2.4	173.913	46.339	71.860
2.5	181.159	43.148	66.912
2.6	188.406	40.322	62.529
2.7	195.652	37.778	58.585
2.8	202.899	35.430	54.944
2.9	210.145	33.261	51.579
3.0	217.391	31.304	48.545
3.1	224.638	29.533	45.799
3.2	231.884	27.904	43.273
3.3	239.130	26.383	40.913
3.4	246.377	24.988	38.751
3.450	250.000	24.300	37.683

Table XI (41) Axial Compression Load Capacity of Single Unequal Angle Section

ISA 10075(6)@ 8.0 kg/m

Section	*Area mm²*	$r_{min} = r_{vv}$ *mm*	*Buckling Class y-y*
ISA 10075(6)@ 8.0 kg/m	*1014.054*	*15.90*	*c*
Eff Length m	*Sl Ratio*	*D.C.S.* f_{cd} *N/mm²*	*F.A.L. kN*
1.0	62.893	163.371	165.667
1.2	75.472	143.245	145.258
1.4	88.050	123.925	125.666
1.6	100.629	106.220	107.713
1.8	113.208	91.104	92.384
2.0	125.786	78.261	79.361
2.2	138.365	67.525	68.473
2.4	150.943	58.643	59.468
2.6	163.522	51.469	52.192
2.8	176.101	45.355	45.992
3.0	188.679	40.215	40.780
3.2	201.258	35.923	36.427
3.4	213.836	32.264	32.718
3.6	226.415	29.125	29.534
3.8	238.994	26.411	26.782
3.975	250.000	24.300	24.642

able XI (42) Axial Compression Load Capacity of Single Unequal Angle Section

ISA 10075(8)@ 10.5 kg/m

Section	Area mm^2	$r_{min} = r_{vv}$ mm	Buckling Class y-y
ISA 10075(8)@ 10.5 kg/m	1336.054	15.90	c
Eff Length m	Sl Ratio	D.C.S. f_{cd} N/mm^2	F.A.L. kN
1.0	62.893	163.371	218.273
1.2	75.472	143.245	191.383
1.4	88.050	123.925	165.570
1.6	100.629	106.220	141.916
1.8	113.208	91.104	121.720
2.0	125.786	78.261	104.561
2.2	138.365	67.525	90.216
2.4	150.943	58.643	78.351
2.6	163.522	51.469	68.765
2.8	176.101	45.355	60.596
3.0	188.679	40.215	53.730
3.2	201.258	35.923	47.995
3.4	213.836	32.264	43.107
3.6	226.415	29.125	38.912
3.8	238.994	26.411	35.287
3.975	250.000	24.300	32.466

able XI (43) Axial Compression Load Capacity of Single Unequal Angle Section

ISA 10075(10)@ 13.0 kg/m

Section	Area mm^2	$r_{min} = r_{vv}$ mm	Buckling Class y-y
ISA 10075(10)@ 13.0 kg/m	1650.054	15.80	c
Eff Length m	Sl Ratio	D.C.S. f_{cd} N/mm^2	F.A.L. kN
1.0	63.291	162.734	268.520
1.2	75.949	142.481	235.101
1.4	88.608	123.089	203.103
1.6	101.266	105.430	173.966
1.8	113.924	90.323	149.037
2.0	126.582	77.513	127.900
2.2	139.241	66.815	110.249
2.4	151.899	58.080	95.835
2.6	164.557	50.930	84.038
2.8	177.215	44.853	74.010
3.0	189.873	39.749	65.589
3.2	202.532	35.541	58.644
3.4	215.190	31.899	52.635
3.6	227.848	28.795	47.513
3.8	240.506	26.104	43.073
3.950	250.000	24.300	40.096

Table XI (44) Axial Compression Load Capacity of Single Unequal Angle Section

ISA 10075(12)@ 15.4 kg/m

Section	Area mm^2	$r_{min} = r_{vv}$ mm	Buckling Class y-y
ISA 10075(12)@ 15.4 kg/m	1956.054	15.80	c
Eff Length m	Sl Ratio	D.C.S. f_{cd} N/mm^2	F.A.L. kN
1.0	63.291	162.734	318.317
1.2	75.949	142.481	278.701
1.4	88.608	123.089	240.768
1.6	101.266	105.430	206.227
1.8	113.924	90.323	176.676
2.0	126.582	77.513	151.619
2.2	139.241	66.815	130.694
2.4	151.899	58.080	113.607
2.6	164.557	50.930	99.623
2.8	177.215	44.853	87.735
3.0	189.873	39.749	77.752
3.2	202.532	35.541	69.519
3.4	215.190	31.899	62.396
3.6	227.848	28.795	56.324
3.8	240.506	26.104	51.060
3.950	250.000	24.300	47.532

Table XI (45) Axial Compression Load Capacity of Single Unequal Angle Section

ISA 12575(6)@ 9.2 kg/m

Section	Area mm^2	$r_{min} = r_{vv}$ mm	Buckling Class y-y
ISA 12575(6)@ 9.2 kg/m	1165.931	16.20	c
Eff Length m	Sl Ratio	D.C.S. f_{cd} N/mm^2	F.A.L. kN
1.0	61.728	165.235	192.652
1.2	74.074	145.481	169.621
1.4	86.420	126.370	147.339
1.6	98.765	108.728	126.770
1.8	111.111	93.389	108.885
2.0	123.457	80.451	93.800
2.2	135.802	69.600	81.149
2.4	148.148	60.496	70.535
2.6	160.494	53.043	61.845
2.8	172.840	46.822	54.591
3.0	185.185	41.578	48.477
3.2	197.531	37.140	43.302
3.4	209.877	33.337	38.869
3.6	222.222	30.089	35.082
3.8	234.568	27.341	31.877
4.0	246.914	24.886	29.016
4.050	250.000	24.300	28.332

able XI (46) Axial Compression Load Capacity of Single Unequal Angle Section

ISA 12575(8)@ 12.1 kg/m

Section	Area mm^2	$r_{min} = r_{vv}$ mm	Buckling Class y-y
ISA 12575(8)@ 12.1 kg/m	1537.931	16.10	c
Eff Length m	Sl Ratio	D.C.S. f_{cd} N/mm²	F.A.L. kN
1.0	62.112	164.621	253.176
1.2	74.534	144.745	222.608
1.4	86.957	125.565	193.111
1.6	99.379	107.870	165.896
1.8	111.801	92.637	142.469
2.0	124.224	79.730	122.619
2.2	136.646	68.917	105.989
2.4	149.068	59.852	92.049
2.6	161.491	52.525	80.780
2.8	173.913	46.339	71.266
3.0	186.335	41.129	63.254
3.2	198.758	36.722	56.476
3.4	211.180	32.981	50.723
3.6	223.602	29.771	45.786
3.8	236.025	27.035	41.578
4.0	248.447	24.595	37.825
4.025	250.000	24.300	37.372

able XI (47) Axial Compression Load Capacity of Single Unequal Angle Section

ISA 12575(10)@ 14.9 kg/m

Section	Area mm^2	$r_{min} = r_{vv}$ mm	Buckling Class y-y
ISA 12575(10)@ 14.9 kg/m	1901.931	16.10	c
Eff Length m	Sl Ratio	D.C.S. f_{cd} N/mm²	F.A.L. kN
1.0	62.112	164.621	313.098
1.2	74.534	144.745	275.296
1.4	86.957	125.565	238.816
1.6	99.379	107.870	205.161
1.8	111.801	92.637	176.189
2.0	124.224	79.730	151.641
2.2	136.646	68.917	131.075
2.4	149.068	59.852	113.835
2.6	161.491	52.525	99.899
2.8	173.913	46.339	88.134
3.0	186.335	41.129	78.225
3.2	198.758	36.722	69.843
3.4	211.180	32.981	62.728
3.6	223.602	29.771	56.623
3.8	236.025	27.035	51.418
4.0	248.447	24.595	46.778
4.025	250.000	24.300	46.217

Table XI (48) Axial Compression Load Capacity of Single Unequal Angle Section

ISA 12595(6)@ 10.1 kg/m

Section	Area mm^2	$r_{min} = r_{vv}$ mm	Buckling Class y-y
ISA 12595(6)@ 10.1 kg/m	1285.931	20.30	c
Eff Length m	Sl Ratio	D.C.S. f_{cd} N/mm^2	F.A.L. kN
1.0	49.261	184.108	236.751
1.2	59.113	169.330	217.747
1.4	68.966	153.655	197.590
1.6	78.818	137.892	177.319
1.8	88.670	122.995	158.163
2.0	98.522	109.069	140.255
2.2	108.374	96.616	124.241
2.4	118.227	85.633	110.118
2.6	128.079	76.106	97.867
2.8	137.931	67.876	87.284
3.0	147.783	60.752	78.123
3.2	157.635	54.695	70.334
3.4	167.488	49.406	63.533
3.6	177.340	44.797	57.606
3.8	187.192	40.795	52.460
4.0	197.044	37.305	47.972
4.2	206.897	34.231	44.019
4.4	216.749	31.478	40.478
4.6	226.601	29.082	37.397
4.8	236.453	26.945	34.649
5.0	246.305	25.002	32.151
5.075	250.000	24.300	31.248

Table XI (49) Axial Compression Load Capacity of Single Unequal Angle Section

ISA 12595(8)@ 13.3 kg/m

Section	Area mm^2	$r_{min} = r_{vv}$ mm	Buckling Class y-y
ISA 12595(8)@ 13.3 kg/m	1697.931	20.20	c
Eff Length m	Sl Ratio	D.C.S. f_{cd} N/mm^2	F.A.L. kN
1.0	49.505	183.743	311.982
1.2	59.406	168.891	286.765
1.4	69.307	153.109	259.968
1.6	79.208	137.267	233.071
1.8	89.109	122.337	207.719
2.0	99.010	108.386	184.032
2.2	108.911	95.950	162.917
2.4	118.812	84.995	144.316
2.6	128.713	75.510	128.211
2.8	138.614	67.323	114.309

Contd...

3.0	148.515	60.240	102.283
3.2	158.416	54.235	92.087
3.4	168.317	48.975	83.157
3.6	178.218	44.402	75.392
3.8	188.119	40.434	68.654
4.0	198.020	36.973	62.778
4.2	207.921	33.924	57.600
4.4	217.822	31.188	52.955
4.6	227.723	28.824	48.941
4.8	237.624	26.699	45.333
5.0	247.525	24.770	42.058
5.050	250.000	24.300	41.260

able XI (50) Axial Compression Load Capacity of Single Unequal Angle Section

ISA 12595(10)@ 16.5 kg/m

Section	*Area mm²*	*$r_{min} = r_{vv}$ mm*	*Buckling Class y-y*
ISA 12595(10)@ 16.5 kg/m	*2101.931*	*20.20*	*c*
Eff Length m	*Sl Ratio*	*D.C.S. f_{cd} N/mm²*	*F.A.L. kN*
1.0	49.505	183.743	386.214
1.2	59.406	168.891	354.997
1.4	69.307	153.109	321.824
1.6	79.208	137.267	288.527
1.8	89.109	122.337	257.143
2.0	99.010	108.386	227.820
2.2	108.911	95.950	201.681
2.4	118.812	84.995	178.654
2.6	128.713	75.510	158.717
2.8	138.614	67.323	141.508
3.0	148.515	60.240	126.620
3.2	158.416	54.235	113.998
3.4	168.317	48.975	102.943
3.6	178.218	44.402	93.330
3.8	188.119	40.434	84.989
4.0	198.020	36.973	77.715
4.2	207.921	33.924	71.305
4.4	217.822	31.188	65.555
4.6	227.723	28.824	60.586
4.8	237.624	26.699	56.119
5.0	247.525	24.770	52.065
5.050	250.000	24.300	51.077

Table XI (51) Axial Compression Load Capacity of Single Unequal Angle Section

ISA 12595(12)@ 19.6 kg/m

Section	Area mm^2	$r_{min} = r_{vv}$ mm	Buckling Class y-y
ISA 12595(12)@ 19.6 kg/m	2497.931	20.10	c
Eff Length m	SI Ratio	D.C.S. f_{cd} N/mm^2	F.A.L. kN
1.0	49.751	183.373	458.054
1.2	59.701	168.448	420.771
1.4	69.652	152.557	381.077
1.6	79.602	136.637	341.309
1.8	89.552	121.672	303.927
2.0	99.502	107.697	269.019
2.2	109.453	95.279	237.999
2.4	119.403	84.351	210.702
2.6	129.353	74.908	187.115
2.8	139.303	66.764	166.772
3.0	149.254	59.722	149.182
3.2	159.204	53.770	134.313
3.4	169.154	48.540	121.249
3.6	179.104	44.003	109.916
3.8	189.055	40.069	100.089
4.0	199.005	36.638	91.520
4.2	208.955	33.613	83.964
4.4	218.905	30.896	77.175
4.6	228.856	28.563	71.349
4.8	238.806	26.451	66.072
5.0	248.756	24.536	61.290
5.025	250.000	24.300	60.700

Table XI (52) Axial Compression Load Capacity of Single Unequal Angle Section

ISA 15075(8)@ 13.7 kg/m

Section	Area mm^2	$r_{min} = r_{vv}$ mm	Buckling Class y-y
ISA 15075(8)@ 13.7 kg/m	1742.009	16.00	c
Eff Length m	SI Ratio	D.C.S. f_{cd} N/mm^2	F.A.L. kN
1.0	62.500	164.000	285.689
1.2	75.000	144.000	250.849
1.4	87.500	124.750	217.316
1.6	100.000	107.000	186.395
1.8	112.500	91.875	160.047
2.0	125.000	79.000	137.619
2.2	137.500	68.225	118.849
2.4	150.000	59.200	103.127
2.6	162.500	52.000	90.584
2.8	175.000	45.850	79.871
3.0	187.500	40.675	70.856
3.2	200.000	36.300	63.235

Contd...

3.4	212.500	32.625	56.833
3.6	225.000	29.450	51.302
3.8	237.500	26.725	46.555
4.000	250.000	24.300	42.331

able XI (53) Axial Compression Load Capacity of Single Unequal Angle Section

ISA 15075(10)@ 16.9 kg/m

Section	Area mm^2	$r_{min} = r_{vv}$ mm	Buckling Class y-y
ISA 15075(10)@ 16.9 kg/m	2156.009	15.90	c
Eff Length m	Sl Ratio	D.C.S. f_{cd} N/mm^2	F.A.L. kN
1.0	62.893	163.371	352.229
1.2	75.472	143.245	308.838
1.4	88.050	123.925	267.182
1.6	100.629	106.220	229.012
1.8	113.208	91.104	196.421
2.0	125.786	78.261	168.731
2.2	138.365	67.525	145.583
2.4	150.943	58.643	126.436
2.6	163.522	51.469	110.967
2.8	176.101	45.355	97.785
3.0	188.679	40.215	86.704
3.2	201.258	35.923	77.450
3.4	213.836	32.264	69.562
3.6	226.415	29.125	62.793
3.8	238.994	26.411	56.943
3.975	250.000	24.300	52.391

able XI (54) Axial Compression Load Capacity of Single Unequal Angle Section

ISA 15075(12)@ 20.1 kg/m

Section	Area mm^2	$r_{min} = r_{vv}$ mm	Buckling Class y-y
ISA 15075(12)@ 20.1 kg/m	2562.009	15.80	c
Eff Length m	Sl Ratio	D.C.S. f_{cd} N/mm^2	F.A.L. kN
1.0	63.291	162.734	416.926
1.2	75.949	142.481	365.038
1.4	88.608	123.089	315.354
1.6	101.266	105.430	270.114
1.8	113.924	90.323	231.408
2.0	126.582	77.513	198.588
2.2	139.241	66.815	171.181
2.4	151.899	58.080	148.801
2.6	164.557	50.930	130.484
2.8	177.215	44.853	114.914
3.0	189.873	39.749	101.838
3.2	202.532	35.541	91.055
3.4	215.190	31.899	81.725

Contd...

3.6	227.848	28.795	73.773
3.8	240.506	26.104	66.878
3.950	250.000	24.300	62.257

Table XI (55) Axial Compression Load Capacity of Single Unequal Angle Section

ISA 150115(8)@ 16.2 kg/m

Section	Area mm^2	$r_{min} = r_{vv}$ mm	Buckling Class y-y
ISA 150115(8)@ 16.2 kg/m	2057.824	24.50	c
Eff Length m	Sl Ratio	D.C.S. f_{cd} N/mm^2	F.A.L. kN
1.0	40.816	196.776	404.929
1.2	48.980	184.531	379.732
1.4	57.143	172.286	354.534
1.6	65.306	159.510	328.244
1.8	73.469	146.449	301.366
2.0	81.633	133.551	274.825
2.2	89.796	121.306	249.627
2.4	97.959	109.857	226.067
2.6	106.122	99.408	204.565
2.8	114.286	89.929	185.057
3.0	122.449	81.398	167.503
3.2	130.612	73.804	151.876
3.4	138.776	67.192	138.269
3.6	146.939	61.343	126.233
3.8	155.102	56.190	115.629
4.0	163.265	51.602	106.188
4.2	171.429	47.457	97.658
4.4	179.592	43.784	90.099
4.6	187.755	40.576	83.497
4.8	195.918	37.688	77.555
5.0	204.082	35.076	72.179
5.2	212.245	32.694	67.278
5.4	220.408	30.506	62.776
5.6	228.571	28.629	58.913
5.8	236.735	26.886	55.326
6.0	244.898	25.269	52.000
6.125	250.000	24.300	50.005

Table XI (56) Axial Compression Load Capacity of Single Unequal Angle Section

ISA 150115(10)@ 20.0 kg/m

Section	Area mm^2	$r_{min} = r_{vv}$ mm	Buckling Class y-y
ISA 150115(10)@ 20.0 kg/m	2551.824	24.40	c
Eff Length m	Sl Ratio	D.C.S. f_{cd} N/mm^2	F.A.L. kN
1.0	40.984	196.525	501.496
1.2	49.180	184.230	470.121
1.4	57.377	171.934	438.746

Contd...

1.6	65.574	159.082	405.949
1.8	73.770	145.967	372.483
2.0	81.967	133.049	339.518
2.2	90.164	120.770	308.185
2.4	98.361	109.295	278.902
2.6	106.557	98.869	252.296
2.8	114.754	89.418	228.179
3.0	122.951	80.926	206.510
3.2	131.148	73.370	187.229
3.4	139.344	66.731	170.286
3.6	147.541	60.921	155.460
3.8	155.738	55.815	142.429
4.0	163.934	51.254	130.791
4.2	172.131	47.141	120.295
4.4	180.328	43.472	110.933
4.6	188.525	40.275	102.776
4.8	196.721	37.415	95.476
5.0	204.918	34.825	88.866
5.2	213.115	32.459	82.830
5.4	221.311	30.298	77.316
5.6	229.508	28.413	72.505
5.8	237.705	26.682	68.088
6.0	245.902	25.079	63.996
6.100	250.000	24.300	62.009

able XI (57) Axial Compression Load Capacity of Single Unequal Angle Section

ISA 150115(12)@ 23.8 kg/m

Section	*Area mm²*	*$r_{min} = r_{vv}$ mm*	*Buckling Class y-y*
ISA 150115(12)@ 23.8 kg/m	*3037.824*	*24.40*	*c*
Eff Length m	*Sl Ratio*	*D.C.S. f_{cd} N/mm²*	*F.A.L. kN*
1.0	40.984	196.525	597.007
1.2	49.180	184.230	559.657
1.4	57.377	171.934	522.307
1.6	65.574	159.082	483.263
1.8	73.770	145.967	443.423
2.0	81.967	133.049	404.180
2.2	90.164	120.770	366.880
2.4	98.361	109.295	332.019
2.6	106.557	98.869	300.346
2.8	114.754	89.418	271.636
3.0	122.951	80.926	245.840
3.2	131.148	73.370	222.887
3.4	139.344	66.731	202.717
3.6	147.541	60.921	185.068
3.8	155.738	55.815	169.555

Contd...

4.0	163.934	51.254	155.701
4.2	172.131	47.141	143.206
4.4	180.328	43.472	132.061
4.6	188.525	40.275	122.350
4.8	196.721	37.415	113.659
5.0	204.918	34.825	105.791
5.2	213.115	32.459	98.605
5.4	221.311	30.298	92.041
5.6	229.508	28.413	86.314
5.8	237.705	26.682	81.055
6.0	245.902	25.079	76.185
6.100	250.000	24.300	73.819

Table XI (58) Axial Compression Load Capacity of Single Unequal Angle Section

ISA 150115(15)@ 29.5 kg/m

Section	*Area mm²*	*$r_{min} = r_{vv}$ mm*	*Buckling Class y-y*
ISA 150115(15)@ 29.5 kg/m	*3751.824*	*24.30*	*c*
Eff Length m	*Sl Ratio*	*D.C.S. f_{cd} N/mm²*	*F.A.L. kN*
1.0	41.152	196.272	736.377
1.2	49.383	183.926	690.058
1.4	57.613	171.580	643.739
1.6	65.844	158.650	595.228
1.8	74.074	145.481	545.821
2.0	82.305	132.543	497.279
2.2	90.535	120.251	451.161
2.4	98.765	108.728	407.930
2.6	106.996	98.325	368.898
2.8	115.226	88.903	333.550
3.0	123.457	80.451	301.837
3.2	131.687	72.933	273.633
3.4	139.918	66.267	248.621
3.6	148.148	60.496	226.971
3.8	156.379	55.437	207.988
4.0	164.609	50.903	190.980
4.2	172.840	46.822	175.669
4.4	181.070	43.183	162.014
4.6	189.300	39.973	149.971
4.8	197.531	37.140	139.341
5.0	205.761	34.572	129.707
5.2	213.992	32.222	120.892
5.4	222.222	30.089	112.888
5.6	230.453	28.205	105.820
5.8	238.683	26.477	99.335
6.0	246.914	24.886	93.369
6.075	250.000	24.300	91.169

able XI (59) Axial Compression Load Capacity of Single Unequal Angle Section

ISA 200100(10)@ 22.8 kg/m

Section	Area mm^2	$r_{min} = r_{vv}$ mm	Buckling Class y-y
ISA 200100(10)@ 22.8 kg/m	2903.434	21.40	c
Eff Length m	Sl Ratio	D.C.S. f_{cd} N/mm^2	F.A.L. kN
1.0	46.729	187.907	545.574
1.2	56.075	173.888	504.872
1.4	65.421	159.327	462.596
1.6	74.766	144.374	419.180
1.8	84.112	129.832	376.958
2.0	93.458	116.159	337.260
2.2	102.804	103.523	300.573
2.4	112.150	92.257	267.862
2.6	121.495	82.294	238.936
2.8	130.841	73.619	213.747
3.0	140.187	66.069	191.827
3.2	149.533	59.527	172.833
3.4	158.879	53.962	156.674
3.6	168.224	49.023	142.336
3.8	177.570	44.693	129.764
4.0	186.916	40.903	118.759
4.2	196.262	37.571	109.085
4.4	205.607	34.618	100.510
4.6	214.953	31.963	92.801
4.8	224.299	29.611	85.974
5.0	233.645	27.535	79.945
5.2	242.991	25.632	74.420
5.350	250.000	24.300	70.553

able XI (60) Axial Compression Load Capacity of Single Unequal Angle Section

ISA 200100(12)@ 27.2 kg/m

Section	Area mm^2	$r_{min} = r_{vv}$ mm	Buckling Class y-y
ISA 200100(12)@ 27.2 kg/m	3459.434	21.30	c
Eff Length m	Sl Ratio	D.C.S. f_{cd} N/mm^2	F.A.L. kN
1.0	46.948	187.577	648.912
1.2	56.338	173.493	600.187
1.4	65.728	158.836	549.481
1.6	75.117	143.812	497.509
1.8	84.507	129.239	447.095
2.0	93.897	115.545	399.719
2.2	103.286	102.925	356.062
2.4	112.676	91.683	317.172
2.6	122.066	81.758	282.837
2.8	131.455	73.121	252.958

Contd...

3.0	140.845	65.608	226.968
3.2	150.235	59.062	204.319
3.4	159.624	53.522	185.154
3.6	169.014	48.613	168.172
3.8	178.404	44.318	153.316
4.0	187.793	40.561	140.317
4.2	197.183	37.258	128.891
4.4	206.573	34.328	118.756
4.6	215.962	31.690	109.630
4.8	225.352	29.369	101.600
5.0	234.742	27.304	94.457
5.2	244.131	25.415	87.922
5.325	250.000	24.300	84.064

Table XI (61) Axial Compression Load Capacity of Single Unequal Angle Section

ISA 200100(15)@ 33.6 kg/m

Section	*Area mm²*	*$r_{min} = r_{vv}$ mm*	*Buckling Class y-y*
ISA 200100(15)@ 33.6 kg/m	*4278.434*	*21.20*	*c*
Eff Length m	*Sl Ratio*	*D.C.S. f_{cd} N/mm²*	*F.A.L. kN*
1.0	47.170	187.245	801.117
1.2	56.604	173.094	740.573
1.4	66.038	158.340	677.446
1.6	75.472	143.245	612.865
1.8	84.906	128.642	550.384
2.0	94.340	114.925	491.697
2.2	103.774	102.321	437.773
2.4	113.208	91.104	389.781
2.6	122.642	81.217	347.481
2.8	132.075	72.619	310.695
3.0	141.509	65.143	278.712
3.2	150.943	58.643	250.902
3.4	160.377	53.104	227.201
3.6	169.811	48.198	206.212
3.8	179.245	43.940	187.993
4.0	188.679	40.215	172.058
4.2	198.113	36.942	158.052
4.4	207.547	34.036	145.620
4.6	216.981	31.415	134.407
4.8	226.415	29.125	124.607
5.0	235.849	27.072	115.824
5.2	245.283	25.196	107.800
5.300	250.000	24.300	103.966

able XI (62) Axial Compression Load Capacity of Single Unequal Angle Section

ISA 200150(10)@ 26.7 kg/m

Section	Area mm^2	$r_{min} = r_{vv}$ mm	Buckling Class y-y
ISA 200150(10)@ 26.7 kg/m	3400.376	32.10	c
Eff Length m	Sl Ratio	D.C.S. f_{cd} N/mm^2	F.A.L. kN
1.0	31.153	209.502	712.384
1.2	37.383	201.402	684.842
1.4	43.614	192.579	654.842
1.6	49.844	183.234	623.063
1.8	56.075	173.888	591.284
2.0	62.305	164.312	558.721
2.2	68.536	154.343	524.823
2.4	74.766	144.374	490.925
2.6	80.997	134.505	457.366
2.8	87.227	125.159	425.587
3.0	93.458	116.159	394.984
3.2	99.688	107.436	365.323
3.4	105.919	99.660	338.883
3.6	112.150	92.257	313.708
3.8	118.380	85.466	290.616
4.0	124.611	79.366	269.874
4.2	130.841	73.619	250.331
4.4	137.072	68.572	233.170
4.6	143.302	63.888	217.245
4.8	149.533	59.527	202.415
5.0	155.763	55.800	189.740
5.2	161.994	52.263	177.715
5.4	168.224	49.023	166.698
5.6	174.455	46.095	156.741
5.8	180.685	43.333	147.347
6.0	186.916	40.903	139.085
6.2	193.146	38.630	131.357
6.4	199.377	36.512	124.154
6.6	205.607	34.618	117.713
6.8	211.838	32.804	111.545
7.0	218.069	31.121	105.825
7.2	224.299	29.611	100.689
7.4	230.530	28.189	95.852
7.6	236.760	26.880	91.403
7.8	242.991	25.632	87.158
8.0	249.221	24.448	83.132
8.025	250.000	24.300	82.629

Table XI (63) Axial Compression Load Capacity of Single Unequal Angle Section

ISA 200150(12)@ 31.8 kg/m

Section	Area mm²	$r_{min} = r_{vv}$ mm	Buckling Class y-y
ISA 200150(12)@ 31.8 kg/m	4056.376	32.10	c
Eff Length m	Sl Ratio	D.C.S. f_{cd} N/mm²	F.A.L. kN
1.0	31.153	209.502	849.817
1.2	37.383	201.402	816.962
1.4	43.614	192.579	781.175
1.6	49.844	183.234	743.264
1.8	56.075	173.888	705.354
2.0	62.305	164.312	666.509
2.2	68.536	154.343	626.072
2.4	74.766	144.374	585.634
2.6	80.997	134.505	545.601
2.8	87.227	125.159	507.691
3.0	93.458	116.159	471.184
3.2	99.688	107.436	435.801
3.4	105.919	99.660	404.260
3.6	112.150	92.257	374.229
3.8	118.380	85.466	346.681
4.0	124.611	79.366	321.938
4.2	130.841	73.619	298.625
4.4	137.072	68.572	278.154
4.6	143.302	63.888	259.156
4.8	149.533	59.527	241.464
5.0	155.763	55.800	226.344
5.2	161.994	52.263	211.999
5.4	168.224	49.023	198.857
5.6	174.455	46.095	186.980
5.8	180.685	43.333	175.774
6.0	186.916	40.903	165.917
6.2	193.146	38.630	156.699
6.4	199.377	36.512	148.106
6.6	205.607	34.618	140.423
6.8	211.838	32.804	133.064
7.0	218.069	31.121	126.240
7.2	224.299	29.611	120.114
7.4	230.530	28.189	114.344
7.6	236.760	26.880	109.037
7.8	242.991	25.632	103.972
8.0	249.221	24.448	99.170
8.025	250.000	24.300	98.570

able XI (64) Axial Compression Load Capacity of Single Unequal Angle Section

ISA 200150(15)@ 39.4 kg/m

Section	Area mm²	$r_{min} = r_{vv}$ mm	Buckling Class y-y
ISA 200150(15)@ 39.4 kg/m	5025.376	32.00	c
Eff Length m	Sl Ratio	D.C.S. f_{cd} N/mm²	F.A.L. kN
1.0	31.250	209.375	1052.188
1.2	37.500	201.250	1011.357
1.4	43.750	192.375	966.757
1.6	50.000	183.000	919.644
1.8	56.250	173.625	872.531
2.0	62.500	164.000	824.162
2.2	68.750	154.000	773.908
2.4	75.000	144.000	723.654
2.6	81.250	134.125	674.028
2.8	87.500	124.750	626.916
3.0	93.750	115.750	581.687
3.2	100.000	107.000	537.715
3.4	106.250	99.250	498.769
3.6	112.500	91.875	461.706
3.8	118.750	85.063	427.471
4.0	125.000	79.000	397.005
4.2	131.250	73.288	368.297
4.4	137.500	68.225	342.856
4.6	143.750	63.575	319.488
4.8	150.000	59.200	297.502
5.0	156.250	55.513	278.971
5.2	162.500	52.000	261.320
5.4	168.750	48.750	244.987
5.6	175.000	45.850	230.413
5.8	181.250	43.113	216.657
6.0	187.500	40.675	204.407
6.2	193.750	38.425	193.100
6.4	200.000	36.300	182.421
6.6	206.250	34.425	172.999
6.8	212.500	32.625	163.953
7.0	218.750	30.938	155.473
7.2	225.000	29.450	147.997
7.4	231.250	28.038	140.899
7.6	237.500	26.725	134.303
7.8	243.750	25.488	128.084
8.000	250.000	24.300	122.117

Table XI (65) Axial Compression Load Capacity of Single Unequal Angle Section

ISA 200150(18)@ 46.9 kg/m

Section	*Area mm²*	$r_{min} = r_{vv}$ *mm*	*Buckling Class y-y*
ISA 200150(18)@ 46.9 kg/m	*5976.376*	*31.90*	*c*
Eff Length m	*Sl Ratio*	*D.C.S.* f_{cd} *N/mm²*	*F.A.L. kN*
1.0	31.348	209.248	1250.543
1.2	37.618	201.097	1201.832
1.4	43.887	192.169	1148.476
1.6	50.157	182.765	1092.272
1.8	56.426	173.361	1036.067
2.0	62.696	163.687	978.252
2.2	68.966	153.655	918.301
2.4	75.235	143.624	858.350
2.6	81.505	133.743	799.298
2.8	87.774	124.339	743.094
3.0	94.044	115.339	689.307
3.2	100.313	106.611	637.149
3.4	106.583	98.837	590.687
3.6	112.853	91.491	546.782
3.8	119.122	84.657	505.940
4.0	125.392	78.632	469.932
4.2	131.661	72.954	436.002
4.4	137.931	67.876	405.652
4.6	144.201	63.260	378.063
4.8	150.470	58.923	352.143
5.0	156.740	55.224	330.036
5.2	163.009	51.735	309.188
5.4	169.279	48.475	289.704
5.6	175.549	45.603	272.541
5.8	181.818	42.891	256.332
6.0	188.088	40.446	241.719
6.2	194.357	38.218	228.408
6.4	200.627	36.112	215.818
6.6	206.897	34.231	204.578
6.8	213.166	32.445	193.904
7.0	219.436	30.752	183.788
7.2	225.705	29.288	175.035
7.4	231.975	27.885	166.653
7.6	238.245	26.569	158.784
7.8	244.514	25.342	151.455
7.975	250.000	24.300	145.226

Factored Axial Compression Load Capacity of Single I-Sections

Table XII (01) Axial Compression Load Capacity of I-Section

ISJB 150 @7.1 kg/m

Section	Area mm²	$r_{min} = r_{yy}$ mm	Buckling Class y-y
ISJB 150 @7.1 kg/m	900.768	10.10	b
Eff Length m	Sl Ratio	D.C.S. f_{cd} N/mm²	F.A.L. kN
1.0	99.010	119.584	107.718
1.2	118.812	93.161	83.917
1.4	138.614	73.075	65.824
1.6	158.416	58.361	52.570
1.8	178.218	47.391	42.688
2.0	198.020	39.233	35.340
2.2	217.822	32.932	29.664
2.4	237.624	28.047	25.263
2.5250	250.000	25.500	22.970

Table XII (02) Axial Compression Load Capacity of I-Section

ISJB175 @ 8.1kg/m

Section	Area mm²	$r_{min} = r_{yy}$ mm	Buckling Class y-y
ISJB175 @ 8.1kg/m	1027.648	9.7	b
Eff Length m	Sl Ratio	D.C.S. f_{cd} N/mm²	F.A.L. kN
1.0	103.093	113.670	116.813
1.2	123.711	87.729	90.154
1.4	144.330	68.423	70.314
1.6	164.948	54.430	55.935
1.8	185.567	44.106	45.326
2.0	206.186	36.459	37.467
2.2	226.804	30.599	31.445
2.4	247.423	26.015	26.735
2.4250	250.000	25.500	26.205

Table XII (03) Axial Compression Load Capacity of I-Section

ISJB200 @9.9kg/m

Section	Area mm²	$r_{min} = r_{yy}$ mm	Buckling Class y-y
ISJB200 @9.9kg/m	1264.368	11.7	b
Eff Length m	Sl Ratio	D.C.S. f_{cd} N/mm²	F.A.L. kN
1.0	85.470	141.248	178.589
1.2	102.564	114.410	144.657
1.4	119.658	92.121	116.474
1.6	136.752	74.788	94.560
1.8	153.846	61.423	77.661
2.0	170.940	51.030	64.521
2.2	188.034	43.045	54.425
2.4	205.128	36.808	46.538

Contd...

2.6	222.222	31.744	40.137
2.8	239.316	27.657	34.969
2.9250	250.000	25.500	32.241

able XII (04) Axial Compression Load Capacity of I-Section

ISJB225 @12.8kg/m

Section	*Area mm²*	$r_{min} = r_{yy}$ *mm*	*Buckling Class y-y*
ISJB225 @12.8kg/m	*1627.796*	*15.80*	*b*
Eff Length m	*Sl Ratio*	*D.C.S.* f_{cd} *N/mm²*	*F.A.L. kN*
1.0	63.291	176.063	286.595
1.2	75.949	156.481	254.719
1.4	88.608	136.228	221.751
1.6	101.266	116.228	189.195
1.8	113.924	99.173	161.434
2.0	126.582	84.657	137.804
2.2	139.241	72.499	118.013
2.4	151.899	62.728	102.108
2.6	164.557	54.657	88.970
2.8	177.215	47.892	77.959
3.0	189.873	42.254	68.782
3.2	202.532	37.665	61.310
3.4	215.190	33.695	54.848
3.6	227.848	30.338	49.384
3.8	240.506	27.399	44.600
3.9500	250.000	25.500	41.509

able XII (05) Axial Compression Load Capacity of I-Section

ISLB75 @6.1kg/m

Section	*Area mm²*	$r_{min} = r_{yy}$ *mm*	*Buckling Class y-y*
ISLB75 @6.1kg/m	*771.383*	*11.4*	*b*
Eff Length m	*Sl Ratio*	*D.C.S.* f_{cd} *N/mm²*	*F.A.L. kN*
1.0	87.719	137.649	106.180
1.2	105.263	110.632	85.339
1.4	122.807	88.696	68.419
1.6	140.351	71.526	55.174
1.8	157.895	58.711	45.288
2.0	175.439	48.781	37.629
2.2	192.982	41.096	31.701
2.4	210.526	35.047	27.035
2.6	228.070	30.282	23.359
2.8	245.614	26.377	20.347
2.8500	250.000	25.500	19.670

Table XII (06) Axial Compression Load Capacity of I-Section

ISLB100 @8.0kg/m

Section	Area mm^2	$r_{min} = r_{yy}$ mm	Buckling Class y-y
ISLB100 @8.0kg/m	1021.096	11.2	b
Eff Length m	Sl Ratio	D.C.S. f_{cd} N/mm^2	F.A.L. kN
1.0	89.286	135.143	137.994
1.2	107.143	108.000	110.278
1.4	125.000	86.350	88.172
1.6	142.857	69.571	71.039
1.8	160.714	56.886	58.086
2.0	178.571	47.214	48.210
2.2	196.429	39.821	40.661
2.4	214.286	33.957	34.673
2.6	232.143	29.307	29.925
2.8000	250.000	25.500	26.038

Table XII (07) Axial Compression Load Capacity of I-Section

ISLB125 @11.9kg/m

Table XII	Area mm^2	$r_{min} = r_{yy}$ mm	Buckling Class y-y
ISLB125 @11.9kg/m	1512.207	16.9	b
Eff Length m	Sl Ratio	D.C.S. f_{cd} N/mm^2	F.A.L. kN
1.0	59.172	182.077	275.338
1.2	71.006	164.391	248.592
1.4	82.840	145.456	219.959
1.6	94.675	126.521	191.325
1.8	106.509	108.888	164.661
2.0	118.343	93.738	141.751
2.2	130.178	80.837	122.242
2.4	142.012	70.231	106.203
2.6	153.846	61.423	92.884
2.8	165.680	54.005	81.667
3.0	177.515	47.743	72.197
3.2	189.349	42.480	64.238
3.4	201.183	38.109	57.629
3.6	213.018	34.325	51.906
3.8	224.852	31.087	47.010
4.0	236.686	28.262	42.738
4.2	248.521	25.796	39.009
4.225	250.000	25.500	38.561

Table XII (08) Axial Compression Load Capacity of I-Section

ISLB150 @ 14.2kg/m

Section	Area mm²	$r_{min} = r_{yy}$ mm	Buckling Class y-y
ISLB150 @ 14.2kg/m	1808.321	17.5	b
Eff Length m	Sl Ratio	D.C.S. f_{cd} N/mm²	F.A.L. kN
1.0	57.143	184.714	334.023
1.2	68.571	168.143	304.056
1.4	80.000	150.000	271.248
1.6	91.429	131.714	238.182
1.8	102.857	114.000	206.149
2.0	114.286	98.729	178.533
2.2	125.714	85.586	154.766
2.4	137.143	74.429	134.591
2.6	148.571	65.114	117.748
2.8	160.000	57.300	103.617
3.0	171.429	50.786	91.837
3.2	182.857	45.271	81.865
3.4	194.286	40.614	73.444
3.6	205.714	36.614	66.210
3.8	217.143	33.129	59.907
4.0	228.571	30.157	54.534
4.2	240.000	27.500	49.729
4.3750	250.000	25.500	46.112

Table XII (09) Axial Compression Load Capacity of I-Section

ISLB175 @16.7kg/m

Section	Area mm²	$r_{min} = r_{yy}$ mm	Buckling Class y-y
ISLB175 @16.7kg/m	2129.721	19.3	b
Eff Length m	Sl Ratio	D.C.S. f_{cd} N/mm²	F.A.L. kN
1.0	51.813	191.642	408.145
1.2	62.176	177.736	378.528
1.4	72.539	161.938	344.882
1.6	82.902	145.358	309.571
1.8	93.264	128.777	274.259
2.0	103.627	112.922	240.493
2.2	113.990	99.093	211.040
2.4	124.352	87.043	185.377
2.6	134.715	76.662	163.269
2.8	145.078	67.839	144.479
3.0	155.440	60.355	128.539
3.2	165.803	53.934	114.865
3.4	176.166	48.417	103.115
3.6	186.528	43.693	93.053
3.8	196.891	39.650	84.444

Contd...

4.0	207.254	36.106	76.896
4.2	217.617	32.991	70.262
4.4	227.979	30.305	64.542
4.6	238.342	27.881	59.379
4.8	248.705	25.759	54.860
4.8250	250.000	25.500	54.308

Table XII (10) Axial Compression Load Capacity of I-Section

ISLB200 @19.8kg/m

Section	*Area mm²*	$r_{min} = r_{yy}$ *mm*	*Buckling Class y-y*
ISLB200 @19.8kg/m	*2526.761*	*21.3*	*b*
Eff Length m	*Sl Ratio*	*D.C.S.* f_{cd} *N/mm²*	*F.A.L. kN*
1.0	46.948	197.662	499.445
1.2	56.338	185.761	469.373
1.4	65.728	172.408	435.635
1.6	75.117	157.812	398.754
1.8	84.507	142.789	360.793
2.0	93.897	127.765	322.832
2.2	103.286	113.399	286.532
2.4	112.676	100.708	254.466
2.6	122.066	89.490	226.119
2.8	131.455	79.661	201.284
3.0	140.845	71.141	179.756
3.2	150.235	63.843	161.315
3.4	159.624	57.552	145.419
3.6	169.014	52.072	131.573
3.8	178.404	47.298	119.511
4.0	187.793	43.149	109.027
4.2	197.183	39.542	99.914
4.4	206.573	36.331	91.800
4.6	215.962	33.471	84.573
4.8	225.352	30.962	78.233
5.0	234.742	28.709	72.542
5.2	244.131	26.674	67.398
5.325	250.000	25.500	64.432

Table XII (11) Axial Compression Load Capacity of I-Section

ISLB225 @23.5kg/m

Section	*Area mm²*	$r_{min} = r_{yy}$ *mm*	*Buckling Class y-y*
ISLB225 @23.5kg/m	*2991.639*	*19.4*	*b*
Eff Length m	*Sl Ratio*	*D.C.S.* f_{cd} *N/mm²*	*F.A.L. kN*
1.0	51.546	191.990	574.364
1.2	61.856	178.216	533.159
1.4	72.165	162.536	486.249

Contd...

1.6	82.474	146.041	436.903
1.8	92.784	129.546	387.556
2.0	103.093	113.670	340.060
2.2	113.402	99.815	298.612
2.4	123.711	87.729	262.453
2.6	134.021	77.301	231.257
2.8	144.330	68.423	204.696
3.0	154.639	60.892	182.166
3.2	164.948	54.430	162.835
3.4	175.258	48.871	146.205
3.6	185.567	44.106	131.950
3.8	195.876	40.026	119.743
4.0	206.186	36.459	109.071
4.2	216.495	33.316	99.671
4.4	226.804	30.599	91.541
4.6	237.113	28.164	84.256
4.8	247.423	26.015	77.829
4.850	250.000	25.500	76.287

able XII (12) Axial Compression Load Capacity of I-Section

ISLB250 @27.9kg/m

Section	*Area mm^2*	*$r_{min} = r_{yy}$ mm*	*Buckling Class y-y*
ISLB250 @27.9kg/m	*3552.887*	*23.3*	*b*
Eff Length m	*Sl Ratio*	*D.C.S. f_{cd} N/mm²*	*F.A.L. kN*
1.000	42.918	202.498	719.452
1.200	51.502	192.047	682.322
1.400	60.086	180.871	642.615
1.600	68.670	167.996	596.870
1.800	77.253	154.395	548.547
2.000	85.837	140.661	499.752
2.200	94.421	126.927	450.957
2.400	103.004	113.794	404.297
2.600	111.588	102.047	362.561
2.800	120.172	91.516	325.147
3.000	128.755	82.332	292.515
3.200	137.339	74.248	263.795
3.400	145.923	67.180	238.684
3.600	154.506	60.981	216.657
3.800	163.090	55.508	197.213
4.000	171.674	50.663	180.000
4.200	180.258	46.389	164.816
4.400	188.841	42.698	151.702
4.600	197.425	39.453	140.171

Contd...

4.800	206.009	36.517	129.741
5.000	214.592	33.868	120.330
5.200	223.176	31.506	111.937
5.400	231.760	29.395	104.438
5.600	240.343	27.431	97.460
5.800	248.927	25.715	91.361
5.825	250.000	25.500	90.599

Table XII (13) Axial Compression Load Capacity of I-Section

ISLB275 @33.0kg/m

Section	*Area mm²*	*$r_{min} = r_{yy}$ mm*	*Buckling Class y-y*
ISLB275 @33.0kg/m	*4201.737*	*26.1*	*b*
Eff Length m	*Sl Ratio*	*D.C.S. f_{cd} N/mm²*	*F.A.L. kN*
1.000	38.314	207.686	872.641
1.200	45.977	198.828	835.421
1.400	53.640	189.268	795.255
1.600	61.303	179.046	752.304
1.800	68.966	167.552	704.008
2.000	76.628	155.395	652.927
2.200	84.291	143.134	601.412
2.400	91.954	130.874	549.896
2.600	99.617	118.613	498.381
2.800	107.280	107.808	452.983
3.000	114.943	97.921	411.437
3.200	122.605	88.912	373.586
3.400	130.268	80.753	339.304
3.600	137.931	73.703	309.682
3.800	145.594	67.437	283.352
4.000	153.257	61.818	259.743
4.200	160.920	56.767	238.519
4.400	168.582	52.322	219.844
4.600	176.245	48.377	203.269
4.800	183.908	44.820	188.320
5.000	191.571	41.619	174.871
5.200	199.234	38.784	162.958
5.400	206.897	36.224	152.204
5.600	214.559	33.878	142.345
5.800	222.222	31.744	133.382
6.000	229.885	29.829	125.332
6.200	237.548	28.064	117.917
6.400	245.211	26.458	111.169
6.525	250.000	25.500	107.144

[T]able XII (14) Axial Compression Load Capacity of I-Section

ISLB300 @37.7kg/m

Section	*Area mm²*	*$r_{min} = r_{yy}$ mm*	*Buckling Class y-y*
ISLB300 @37.7kg/m	*4807.789*	*28*	*b*
Eff Length m	*Sl Ratio*	*D.C.S. f_{cd} N/mm²*	*F.A.L. kN*
1.000	35.714	210.286	1011.009
1.200	42.857	202.571	973.921
1.400	50.000	194.000	932.711
1.600	57.143	184.714	888.067
1.800	64.286	174.571	839.303
2.000	71.429	163.714	787.104
2.200	78.571	152.286	732.158
2.400	85.714	140.857	677.211
2.600	92.857	129.429	622.265
2.800	100.000	118.000	567.319
3.000	107.143	108.000	519.241
3.200	114.286	98.729	474.666
3.400	121.429	90.171	433.525
3.600	128.571	82.529	396.780
3.800	135.714	75.743	364.156
4.000	142.857	69.571	334.485
4.200	150.000	64.000	307.698
4.400	157.143	59.214	284.690
4.600	164.286	54.814	263.536
4.800	171.429	50.786	244.167
5.000	178.571	47.214	226.996
5.200	185.714	44.043	211.749
5.400	192.857	41.143	197.806
5.600	200.000	38.500	185.100
5.800	207.143	36.143	173.767
6.000	214.286	33.957	163.259
6.200	221.429	31.943	153.575
6.400	228.571	30.157	144.989
6.600	235.714	28.486	136.953
6.800	242.857	26.929	129.467
7.000	250.000	25.500	122.599

Table XII (15) Axial Compression Load Capacity of I-Section

ISLB325 @43.1kg/m

Section	Area mm^2	$r_{min} = r_{yy}$ mm	Buckling Class y-y
ISLB325 @43.1kg/m	5489.843	30.5	b
Eff Length m	Sl Ratio	D.C.S. f_{cd} N/mm^2	F.A.L. kN
1.000	32.787	213.213	1170.507
1.200	39.344	206.656	1134.508
1.400	45.902	198.918	1092.029
1.600	52.459	190.803	1047.480
1.800	59.016	182.279	1000.681
2.000	65.574	172.639	947.763
2.200	72.131	162.590	892.594
2.400	78.689	152.098	834.996
2.600	85.246	141.607	777.398
2.800	91.803	131.115	719.799
3.000	98.361	120.623	662.201
3.200	104.918	111.115	610.003
3.400	111.475	102.185	560.981
3.600	118.033	94.120	516.702
3.800	124.590	86.789	476.455
4.000	131.148	79.944	438.881
4.200	137.705	73.911	405.762
4.400	144.262	68.475	375.919
4.600	150.820	63.451	348.335
4.800	157.377	59.057	324.216
5.000	163.934	55.018	302.040
5.200	170.492	51.254	281.377
5.400	177.049	47.975	263.377
5.600	183.607	44.949	246.764
5.800	190.164	42.139	231.338
6.000	196.721	39.713	218.019
6.200	203.279	37.418	205.419
6.400	209.836	35.254	193.539
6.600	216.393	33.346	183.064
6.800	222.951	31.562	173.272
7.000	229.508	29.923	164.272
7.200	236.066	28.405	155.939
7.400	242.623	26.975	148.091
7.600	249.180	25.664	140.891
7.625	250.000	25.500	139.991

able XII (16) Axial Compression Load Capacity of I-Section

ISLB350 @49.5kg/m

Section	*Area mm²*	$r_{min} = r_{yy}$ *mm*	*Buckling Class y-y*
ISLB350 @49.5kg/m	*6301.323*	*31.7*	*b*
Eff Length m	*Sl Ratio*	*D.C.S.* f_{cd} *N/mm²*	*F.A.L. kN*
1.000	31.546	214.454	1351.346
1.200	37.855	208.145	1311.590
1.400	44.164	201.003	1266.586
1.600	50.473	193.385	1218.580
1.800	56.782	185.183	1166.898
2.000	63.091	176.363	1111.319
2.200	69.401	166.899	1051.685
2.400	75.710	156.864	988.453
2.600	82.019	146.770	924.843
2.800	88.328	136.675	861.234
3.000	94.637	126.580	797.624
3.200	100.946	116.675	735.207
3.400	107.256	107.842	679.549
3.600	113.565	99.615	627.709
3.800	119.874	91.855	578.809
4.000	126.183	85.084	536.143
4.200	132.492	78.707	495.960
4.400	138.801	72.903	459.384
4.600	145.110	67.814	427.317
4.800	151.420	63.049	397.291
5.000	157.729	58.822	370.655
5.200	164.038	54.958	346.308
5.400	170.347	51.326	323.425
5.600	176.656	48.172	303.547
5.800	182.965	45.225	284.977
6.000	189.274	42.512	267.882
6.200	195.584	40.134	252.898
6.400	201.893	37.875	238.665
6.600	208.202	35.793	225.546
6.800	214.511	33.892	213.563
7.000	220.820	32.095	202.241
7.200	227.129	30.518	192.302
7.400	233.438	29.009	182.796
7.600	239.748	27.558	173.652
7.800	246.057	26.289	165.653
7.925	250.000	25.500	160.684

Table XII (17) Axial Compression Load Capacity of I-Section

ISLB400 @56.9kg/m

Section	*Area mm²*	*$r_{min} = r_{yy}$ mm*	*Buckling Class y-y*
ISLB400 @56.9kg/m	*7243.043*	*31.5*	*b*
Eff Length m	*Sl Ratio*	*D.C.S. f_{cd} N/mm²*	*F.A.L. kN*
1.000	31.746	214.254	1551.851
1.200	38.095	207.905	1505.863
1.400	44.444	200.667	1453.437
1.600	50.794	192.968	1397.677
1.800	57.143	184.714	1337.894
2.000	63.492	175.762	1273.051
2.200	69.841	166.238	1204.070
2.400	76.190	156.095	1130.605
2.600	82.540	145.937	1057.024
2.800	88.889	135.778	983.444
3.000	95.238	125.619	909.864
3.200	101.587	115.778	838.583
3.400	107.937	106.889	774.201
3.600	114.286	98.729	715.095
3.800	120.635	91.021	659.266
4.000	126.984	84.227	610.060
4.200	133.333	77.933	564.474
4.400	139.683	72.092	522.166
4.600	146.032	67.095	485.974
4.800	152.381	62.405	452.000
5.000	158.730	58.151	421.189
5.200	165.079	54.354	393.688
5.400	171.429	50.786	367.843
5.600	177.778	47.611	344.849
5.800	184.127	44.725	323.948
6.000	190.476	42.024	304.380
6.200	196.825	39.675	287.365
6.400	203.175	37.452	271.269
6.600	209.524	35.357	256.093
6.800	215.873	33.497	242.619
7.000	222.222	31.744	229.926
7.200	228.571	30.157	218.429
7.400	234.921	28.668	207.645
7.600	241.270	27.246	197.344
7.800	247.619	25.976	188.147
7.875	250.000	25.500	184.698

able XII (18) Axial Compression Load Capacity of I-Section

ISLB450 @65.3kg/m

Section	Area mm^2	$r_{min} = r_{yy}$ mm	Buckling Class y-y
ISLB450 @65.3kg/m	8313.563	32	b
Eff Length m	Sl Ratio	D.C.S. f_{cd} N/mm^2	F.A.L. kN
1.000	31.250	214.750	1785.338
1.200	37.500	208.500	1733.378
1.400	43.750	201.500	1675.183
1.600	50.000	194.000	1612.831
1.800	56.250	185.875	1545.284
2.000	62.500	177.250	1473.579
2.200	68.750	167.875	1395.639
2.400	75.000	158.000	1313.543
2.600	81.250	148.000	1230.407
2.800	87.500	138.000	1147.272
3.000	93.750	128.000	1064.136
3.200	100.000	118.000	981.000
3.400	106.250	109.250	908.257
3.600	112.500	100.925	839.046
3.800	118.750	93.238	775.136
4.000	125.000	86.350	717.876
4.200	131.250	79.850	663.838
4.400	137.500	74.100	616.035
4.600	143.750	68.875	572.597
4.800	150.000	64.000	532.068
5.000	156.250	59.813	497.255
5.200	162.500	55.850	464.312
5.400	168.750	52.225	434.176
5.600	175.000	49.000	407.365
5.800	181.250	45.963	382.112
6.000	187.500	43.275	359.769
6.200	193.750	40.813	339.297
6.400	200.000	38.500	320.072
6.600	206.250	36.438	302.925
6.800	212.500	34.475	286.610
7.000	218.750	32.663	271.542
7.200	225.000	31.050	258.136
7.400	231.250	29.513	245.354
7.600	237.500	28.075	233.403
7.800	243.750	26.750	222.388
8.000	250.000	25.500	211.996

Table XII (19) Axial Compression Load Capacity of I-Section

ISLB500 @75.0kg/m

Section	Area mm^2	$r_{min} = r_{yy}$ mm	Buckling Class y-y
ISLB500 @75.0kg/m	9549.819	33.4	b
Eff Length m	Sl Ratio	D.C.S. f_{cd} N/mm^2	F.A.L. kN
1.000	29.940	216.054	2063.276
1.200	35.928	210.072	2006.148
1.400	41.916	203.701	1945.304
1.600	47.904	196.515	1876.682
1.800	53.892	188.940	1804.344
2.000	59.880	181.156	1730.004
2.200	65.868	172.198	1644.456
2.400	71.856	163.030	1556.906
2.600	77.844	153.449	1465.411
2.800	83.832	143.868	1373.916
3.000	89.820	134.287	1282.421
3.200	95.808	124.707	1190.925
3.400	101.796	115.485	1102.861
3.600	107.784	107.102	1022.803
3.800	113.772	99.360	948.869
4.000	119.760	91.995	878.532
4.200	125.749	85.549	816.978
4.400	131.737	79.402	758.279
4.600	137.725	73.893	705.669
4.800	143.713	68.904	658.023
5.000	149.701	64.234	613.419
5.200	155.689	60.189	574.790
5.400	161.677	56.328	537.918
5.600	167.665	52.854	504.751
5.800	173.653	49.674	474.374
6.000	179.641	46.680	445.782
6.200	185.629	44.080	420.953
6.400	191.617	41.602	397.290
6.600	197.605	39.386	376.131
6.800	203.593	37.314	356.346
7.000	209.581	35.338	337.475
7.200	215.569	33.585	320.731
7.400	221.557	31.911	304.742
7.600	227.545	30.414	290.446
7.800	233.533	28.987	276.825
8.000	239.521	27.610	263.672
8.200	245.509	26.398	252.098
8.350	250.000	25.500	243.520

able XII (20) Axial Compression Load Capacity of I-Section

ISLB550 @86.3kg/m

Section	Area mm²	$r_{min} = r_{yy}$ mm	Buckling Class y-y
ISLB550 @86.3kg/m	10997.398	34.8	b
Eff Length m	Sl Ratio	D.C.S. f_{cd} N/mm²	F.A.L. kN
1.000	28.736	217.138	2387.952
1.200	34.483	211.517	2326.139
1.400	40.230	205.724	2262.430
1.600	45.977	198.828	2186.586
1.800	51.724	191.759	2108.846
2.000	57.471	184.287	2026.681
2.200	63.218	176.172	1937.438
2.400	68.966	167.552	1842.633
2.600	74.713	158.460	1742.645
2.800	80.460	149.264	1641.520
3.000	86.207	140.069	1540.394
3.200	91.954	130.874	1439.269
3.400	97.701	121.678	1338.143
3.600	103.448	113.172	1244.602
3.800	109.195	105.126	1156.117
4.000	114.943	97.921	1076.873
4.200	120.690	90.962	1000.346
4.400	126.437	84.813	932.718
4.600	132.184	78.991	868.693
4.800	137.931	73.703	810.546
5.000	143.678	68.931	758.062
5.200	149.425	64.448	708.763
5.400	155.172	60.534	665.722
5.600	160.920	56.767	624.286
5.800	166.667	53.433	587.628
6.000	172.414	50.293	553.093
6.200	178.161	47.420	521.492
6.400	183.908	44.820	492.898
6.600	189.655	42.348	465.721
6.800	195.402	40.201	442.108
7.000	201.149	38.121	419.228
7.200	206.897	36.224	398.371
7.400	212.644	34.433	378.677
7.600	218.391	32.767	360.348
7.800	224.138	31.266	343.839
8.000	229.885	29.829	328.038
8.200	235.632	28.505	313.476
8.400	241.379	27.224	299.395
8.600	247.126	26.075	286.754
8.700	250.000	25.500	280.434

Table XII (21) Axial Compression Load Capacity of I-Section

ISLB600 @99.5kg/m

Section	Area mm^2	$r_{min} = r_{yy}$ mm	Buckling Class y-y
ISLB600 @99.5kg/m	12668.942	37.9	b
Eff Length m	Sl Ratio	D.C.S. f_{cd} N/mm^2	F.A.L. kN
1.000	26.385	219.253	2777.707
1.200	31.662	214.338	2715.432
1.400	36.939	209.061	2648.578
1.600	42.216	203.340	2576.107
1.800	47.493	197.008	2495.882
2.000	52.770	190.398	2412.147
2.200	58.047	183.538	2325.236
2.400	63.325	176.013	2229.901
2.600	68.602	168.098	2129.619
2.800	73.879	159.794	2024.423
3.000	79.156	151.351	1917.456
3.200	84.433	142.908	1810.489
3.400	89.710	134.464	1703.521
3.600	94.987	126.021	1596.554
3.800	100.264	117.631	1490.255
4.000	105.541	110.243	1396.659
4.200	110.818	102.994	1304.824
4.400	116.095	96.503	1222.593
4.600	121.372	90.232	1143.143
4.800	126.649	84.585	1071.609
5.000	131.926	79.228	1003.735
5.200	137.203	74.373	942.228
5.400	142.480	69.865	885.121
5.600	147.757	65.749	832.975
5.800	153.034	61.967	785.057
6.000	158.311	58.431	740.264
6.200	163.588	55.219	699.563
6.400	168.865	52.158	660.787
6.600	174.142	49.429	626.210
6.800	179.420	46.790	592.783
7.000	184.697	44.480	563.521
7.200	189.974	42.211	534.773
7.400	195.251	40.257	510.017
7.600	200.528	38.326	485.548
7.800	205.805	36.584	463.486
8.000	211.082	34.886	441.972
8.200	216.359	33.356	422.584
8.400	221.636	31.891	404.026
8.600	226.913	30.572	387.312
8.800	232.190	29.296	371.153

Contd...

9.000	237.467	28.083	355.777
9.200	242.744	26.951	341.443
9.400	248.021	25.896	328.072
9.475	250.000	25.500	323.058

able XII (22) Axial Compression Load Capacity of I-Section

ISMB100 @11.5kg/m

Section	Area mm^2	$r_{min} = r_{yy}$ mm	Buckling Class y-y
ISMB100 @11.5kg/m	1459.750	16.7	b
Eff Length m	Sl Ratio	D.C.S. f_{cd} N/mm^2	F.A.L. kN
1.000	59.880	181.156	264.442
1.200	71.856	163.030	237.983
1.400	83.832	143.868	210.012
1.600	95.808	124.707	182.040
1.800	107.784	107.102	156.342
2.000	119.760	91.995	134.289
2.200	131.737	79.402	115.908
2.400	143.713	68.904	100.583
2.600	155.689	60.189	87.860
2.800	167.665	52.854	77.154
3.000	179.641	46.680	68.141
3.200	191.617	41.602	60.728
3.400	203.593	37.314	54.470
3.600	215.569	33.585	49.026
3.800	227.545	30.414	44.396
4.000	239.521	27.610	40.304
4.175	250.000	25.500	37.224

able XII (23) Axial Compression Load Capacity of I-Section

ISMB125 @13.0kg/m

Section	Area mm^2	$r_{min} = r_{yy}$ mm	Buckling Class y-y
ISMB125 @13.0kg/m	1660.470	16.2	b
Eff Length m	Sl Ratio	D.C.S. f_{cd} N/mm^2	F.A.L. kN
1.000	61.728	178.407	296.240
1.200	74.074	159.481	264.814
1.400	86.420	139.728	232.015
1.600	98.765	119.975	199.215
1.800	111.111	102.633	170.420
2.000	123.457	88.001	146.123
2.200	135.802	75.662	125.634
2.400	148.148	65.444	108.669
2.600	160.494	57.014	94.669
2.800	172.840	50.080	83.157
3.000	185.185	44.270	73.510
3.200	197.531	39.414	65.445

Contd...

3.400	209.877	35.241	58.516
3.600	222.222	31.744	52.711
3.800	234.568	28.749	47.737
4.000	246.914	26.117	43.367
4.050	250.000	25.500	42.342

Table XII (24) Axial Compression Load Capacity of I-Section

ISMB150 @14.9kg/m

Section	*Area mm²*	$r_{min} = r_{yy}$ *mm*	*Buckling Class y-y*
ISMB150 @14.9kg/m	*1900.390*	*16.6*	*b*
Eff Length m	*Sl Ratio*	*D.C.S.* f_{cd} *N/mm²*	*F.A.L. kN*
1.000	60.241	180.639	343.284
1.200	72.289	162.337	308.504
1.400	84.337	143.060	271.870
1.600	96.386	123.783	235.236
1.800	108.434	106.193	201.808
2.000	120.482	91.184	173.286
2.200	132.530	78.672	149.508
2.400	144.578	68.229	129.662
2.600	156.627	59.560	113.188
2.800	168.675	52.269	99.331
3.000	180.723	46.189	87.777
3.200	192.771	41.175	78.248
3.400	204.819	36.910	70.143
3.600	216.867	33.208	63.109
3.800	228.916	30.071	57.147
4.000	240.964	27.307	51.894
4.150	250.000	25.500	48.460

Table XII (25) Axial Compression Load Capacity of I-Section

ISMB175 @19.3kg/m

Section	*Area mm²*	$r_{min} = r_{yy}$ *mm*	*Buckling Class y-y*
ISMB175 @19.3kg/m	*2462.011*	*18.6*	*b*
Eff Length m	*Sl Ratio*	*D.C.S.* f_{cd} *N/mm²*	*F.A.L. kN*
1.000	53.763	189.108	465.585
1.200	64.516	174.226	428.946
1.400	75.269	157.570	387.939
1.600	86.022	140.366	345.582
1.800	96.774	123.161	303.224
2.000	107.527	107.462	264.573
2.200	118.280	93.816	230.976
2.400	129.032	82.035	201.972
2.600	139.785	71.998	177.259

Contd...

2.800	150.538	63.640	156.682
3.000	161.290	56.552	139.231
3.200	172.043	50.478	124.279
3.400	182.796	45.298	111.524
3.600	193.548	40.887	100.664
3.800	204.301	37.081	91.293
4.000	215.054	33.734	83.054
4.200	225.806	30.848	75.949
4.400	236.559	28.291	69.654
4.600	247.312	26.038	64.105
4.650	250.000	25.500	62.781

able XII (26) Axial Compression Load Capacity of I-Section

ISMB200 @25.4kg/m

Section	*Area mm²*	$r_{min} = r_{yy}$ *mm*	*Buckling Class y-y*
ISMB200 @25.4kg/m	*3232.674*	*21.5*	*b*
Eff Length m	*Sl Ratio*	*D.C.S.* f_{cd} *N/mm²*	*F.A.L. kN*
1.000	46.512	198.186	640.671
1.200	55.814	186.442	602.706
1.400	65.116	173.326	560.305
1.600	74.419	158.930	513.770
1.800	83.721	144.047	465.655
2.000	93.023	129.163	417.541
2.200	102.326	114.744	370.931
2.400	111.628	101.998	329.725
2.600	120.930	90.705	293.219
2.800	130.233	80.786	261.155
3.000	139.535	72.228	233.489
3.200	148.837	64.907	209.823
3.400	158.140	58.547	189.262
3.600	167.442	52.984	171.279
3.800	176.744	48.128	155.582
4.000	186.047	43.900	141.914
4.200	195.349	40.221	130.021
4.400	204.651	36.965	119.496
4.600	213.953	34.053	110.084
4.800	223.256	31.486	101.784
5.000	232.558	29.212	94.432
5.200	241.860	27.128	87.696
5.375	250.000	25.500	82.433

Table XII (27) Axial Compression Load Capacity of I-Section

ISMB225 @31.2kg/m

Section	Area mm²	$r_{min} = r_{yy}$ mm	Buckling Class y-y
ISMB225 @31.2kg/m	3971.499	23.4	b
Eff Length m	Sl Ratio	D.C.S. f_{cd} N/mm²	F.A.L. kN
1.000	42.735	202.718	805.094
1.200	51.282	192.333	763.852
1.400	59.829	181.222	719.724
1.600	68.376	168.436	668.943
1.800	76.923	154.923	615.277
2.000	85.470	141.248	560.966
2.200	94.017	127.573	506.655
2.400	102.564	114.410	454.380
2.600	111.111	102.633	407.608
2.800	119.658	92.121	365.857
3.000	128.205	82.921	329.319
3.200	136.752	74.788	297.021
3.400	145.299	67.667	268.738
3.600	153.846	61.423	243.942
3.800	162.393	55.912	222.054
4.000	170.940	51.030	202.665
4.200	179.487	46.756	185.693
4.400	188.034	43.045	170.954
4.600	196.581	39.765	157.926
4.800	205.128	36.808	146.182
5.000	213.675	34.134	135.564
5.200	222.222	31.744	126.073
5.400	230.769	29.623	117.648
5.600	239.316	27.657	109.841
5.800	247.863	25.927	102.970
5.850	250.000	25.500	101.273

Table XII (28) Axial Compression Load Capacity of I-Section

ISMB250 @37.3kg/m

Section	Area mm²	$r_{min} = r_{yy}$ mm	Buckling Class y-y
ISMB250 @37.3kg/m	4755.427	26.5	b
Eff Length m	Sl Ratio	D.C.S. f_{cd} N/mm²	F.A.L. kN
1.000	37.736	208.264	990.385
1.200	45.283	199.660	949.470
1.400	52.830	190.321	905.056
1.600	60.377	180.434	858.040
1.800	67.925	169.113	804.205
2.000	75.472	157.245	747.768

Contd...

2.200	83.019	145.170	690.344
2.400	90.566	133.094	632.920
2.600	98.113	121.019	575.496
2.800	105.660	110.075	523.456
3.000	113.208	100.055	475.803
3.200	120.755	90.892	432.232
3.400	128.302	82.817	393.830
3.600	135.849	75.619	359.600
3.800	143.396	69.151	328.842
4.000	150.943	63.368	301.342
4.200	158.491	58.311	277.295
4.400	166.038	53.798	255.833
4.600	173.585	49.708	236.381
4.800	181.132	46.013	218.812
5.000	188.679	42.768	203.380
5.200	196.226	39.896	189.724
5.400	203.774	37.255	177.162
5.600	211.321	34.817	165.570
5.800	218.868	32.628	155.162
6.000	226.415	30.696	145.974
6.200	233.962	28.889	137.378
6.400	241.509	27.198	129.339
6.600	249.057	25.689	122.161
6.625	250.000	25.500	121.263

able XII (29) Axial Compression Load Capacity of I-Section

ISMB300 @44.2kg/m

Section	*Area mm²*	$r_{min} = r_{yy}$ *mm*	*Buckling Class y-y*
ISMB300 @44.2kg/m	*5626.377*	*28.4*	*b*
Eff Length m	*Sl Ratio*	*D.C.S.* f_{cd} *N/mm²*	*F.A.L. kN*
1.000	35.211	210.789	1185.977
1.200	42.254	203.296	1143.819
1.400	49.296	194.845	1096.272
1.600	56.338	185.761	1045.159
1.800	63.380	175.930	989.846
2.000	70.423	165.324	930.175
2.200	77.465	154.056	866.779
2.400	84.507	142.789	803.383
2.600	91.549	131.521	739.987
2.800	98.592	120.254	676.592
3.000	105.634	110.113	619.535
3.200	112.676	100.708	566.624
3.400	119.718	92.046	517.888
3.600	126.761	84.466	475.239
3.800	133.803	77.501	436.052

Contd...

4.000	140.845	71.141	400.265
4.200	147.887	65.648	369.360
4.400	154.930	60.697	341.505
4.600	161.972	56.156	315.957
4.800	169.014	52.072	292.976
5.000	176.056	48.472	272.721
5.200	183.099	45.168	254.130
5.400	190.141	42.148	237.140
5.600	197.183	39.542	222.480
5.800	204.225	37.106	208.770
6.000	211.268	34.832	195.980
6.200	218.310	32.790	184.490
6.400	225.352	30.962	174.204
6.600	232.394	29.249	164.568
6.800	239.437	27.630	155.454
7.000	246.479	26.204	147.435
7.100	250.000	25.500	143.473

Table XII (30) Axial Compression Load Capacity of I-Section

ISMB350 @52.4kg/m

Section	*Area mm²*	$r_{min} = r_{yy}$ *mm*	*Buckling Class y-y*
ISMB350 @52.4kg/m	*6671.337*	*28.4*	*b*
Eff Length m	*Sl Ratio*	*D.C.S.* f_{cd} *N/mm²*	*F.A.L. kN*
1.000	35.211	210.789	1406.243
1.200	42.254	203.296	1356.255
1.400	49.296	194.845	1299.877
1.600	56.338	185.761	1239.271
1.800	63.380	175.930	1173.685
2.000	70.423	165.324	1102.932
2.200	77.465	154.056	1027.762
2.400	84.507	142.789	952.592
2.600	91.549	131.521	877.422
2.800	98.592	120.254	802.252
3.000	105.634	110.113	734.599
3.200	112.676	100.708	671.860
3.400	119.718	92.046	614.073
3.600	126.761	84.466	563.502
3.800	133.803	77.501	517.038
4.000	140.845	71.141	474.605
4.200	147.887	65.648	437.959
4.400	154.930	60.697	404.931
4.600	161.972	56.156	374.638
4.800	169.014	52.072	347.389
5.000	176.056	48.472	323.372
5.200	183.099	45.168	301.328
5.400	190.141	42.148	281.183

Contd...

5.600	197.183	39.542	263.800
5.800	204.225	37.106	247.544
6.000	211.268	34.832	232.379
6.200	218.310	32.790	218.754
6.400	225.352	30.962	206.558
6.600	232.394	29.249	195.132
6.800	239.437	27.630	184.326
7.000	246.479	26.204	174.817
7.100	250.000	25.500	170.119

able XII (31) Axial Compression Load Capacity of I-Section

ISMB400 @61.6kg/m

Section	*Area mm²*	$r_{min} = r_{yy}$ *mm*	*Buckling Class y-y*
ISMB400 @61.6kg/m	*7845.577*	*28.2*	*b*
Eff Length m	*Sl Ratio*	*D.C.S.* f_{cd} *N/mm²*	*F.A.L. kN*
1.000	35.461	210.539	1651.800
1.200	42.553	202.936	1592.151
1.400	49.645	194.426	1525.380
1.600	56.738	185.241	1453.324
1.800	63.830	175.255	1374.979
2.000	70.922	164.525	1290.792
2.200	78.014	153.177	1201.764
2.400	85.106	141.830	1112.736
2.600	92.199	130.482	1023.709
2.800	99.291	119.135	934.681
3.000	106.383	109.064	855.669
3.200	113.475	99.726	782.404
3.400	120.567	91.093	714.676
3.600	127.660	83.504	655.139
3.800	134.752	76.628	601.194
4.000	141.844	70.362	552.028
4.200	148.936	64.830	508.627
4.400	156.028	59.961	470.429
4.600	163.121	55.490	435.352
4.800	170.213	51.394	403.213
5.000	177.305	47.848	375.391
5.200	184.397	44.609	349.985
5.400	191.489	41.649	326.760
5.600	198.582	39.025	306.172
5.800	205.674	36.628	287.365
6.000	212.766	34.398	269.871
6.200	219.858	32.341	253.735
6.400	226.950	30.562	239.780

Contd...

6.600	234.043	28.870	226.503
6.800	241.135	27.273	213.973
7.000	248.227	25.855	202.844
7.050	250.000	25.500	200.062

Table XII (32) Axial Compression Load Capacity of I-Section

ISMB450 @72.4kg/m

Section	*Area mm²*	$r_{min} = r_{yy}$ *mm*	*Buckling Class y-y*
ISMB450 @72.4kg/m	*9226.629*	*30.1*	*b*
Eff Length m	*Sl Ratio*	*D.C.S.* f_{cd} *N/mm²*	*F.A.L. kN*
1.000	33.223	212.777	1963.218
1.200	39.867	206.133	1901.912
1.400	46.512	198.186	1828.589
1.600	53.156	189.897	1752.109
1.800	59.801	181.259	1672.411
2.000	66.445	171.332	1580.819
2.200	73.090	161.056	1486.008
2.400	79.734	150.425	1387.918
2.600	86.379	139.794	1289.828
2.800	93.023	129.163	1191.737
3.000	99.668	118.532	1093.647
3.200	106.312	109.163	1007.205
3.400	112.957	100.363	926.013
3.600	119.601	92.190	850.606
3.800	126.246	85.017	784.420
4.000	132.890	78.341	722.822
4.200	139.535	72.228	666.420
4.400	146.179	66.980	618.000
4.600	152.824	62.108	573.047
4.800	159.468	57.656	531.972
5.000	166.113	53.754	495.973
5.200	172.757	50.121	462.450
5.400	179.402	46.799	431.797
5.600	186.047	43.900	405.049
5.800	192.691	41.204	380.177
6.000	199.336	38.746	357.494
6.200	205.980	36.527	337.017
6.400	212.625	34.439	317.755
6.600	219.269	32.512	299.976
6.800	225.914	30.822	284.379
7.000	232.558	29.212	269.525
7.200	239.203	27.683	255.424
7.400	245.847	26.331	242.942
7.525	250.000	25.500	235.279

able XII (33) Axial Compression Load Capacity of I-Section

ISMB500 @86.9kg/m

Section	Area mm^2	$r_{min} = r_{yy}$ mm	Buckling Class y-y
ISMB500 @86.9kg/m	11074.379	35.2	b
Eff Length m	Sl Ratio	D.C.S. f_{cd} N/mm^2	F.A.L. kN
1.000	28.409	217.432	2407.922
1.200	34.091	211.909	2346.762
1.400	39.773	206.227	2283.839
1.600	45.455	199.455	2208.835
1.800	51.136	192.523	2132.070
2.000	56.818	185.136	2050.270
2.200	62.500	177.250	1962.934
2.400	68.182	168.727	1868.550
2.600	73.864	159.818	1769.887
2.800	79.545	150.727	1669.211
3.000	85.227	141.636	1568.535
3.200	90.909	132.545	1467.859
3.400	96.591	123.455	1367.182
3.600	102.273	114.818	1271.540
3.800	107.955	106.864	1183.448
4.000	113.636	99.527	1102.203
4.200	119.318	92.539	1024.808
4.400	125.000	86.350	956.273
4.600	130.682	80.373	890.078
4.800	136.364	75.145	832.189
5.000	142.045	70.205	777.472
5.200	147.727	65.773	728.392
5.400	153.409	61.716	683.465
5.600	159.091	57.909	641.307
5.800	164.773	54.532	603.906
6.000	170.455	51.273	567.814
6.200	176.136	48.432	536.352
6.400	181.818	45.718	506.300
6.600	187.500	43.275	479.244
6.800	193.182	41.023	454.301
7.000	198.864	38.920	431.020
7.200	204.545	37.000	409.752
7.400	210.227	35.134	389.088
7.600	215.909	33.486	370.841
7.800	221.591	31.902	353.298
8.000	227.273	30.482	337.567
8.200	232.955	29.120	322.491
8.400	238.636	27.814	308.019
8.600	244.318	26.636	294.981
8.800	250.000	25.500	282.397

Table XII (34) Axial Compression Load Capacity of I-Section

ISMB550 @103.7kg/m

Section	Area mm²	$r_{min} = r_{yy}$ mm	Buckling Class y-y
ISMB550 @103.7kg/m	13211.078	37.3	b
Eff Length m	Sl Ratio	D.C.S. f_{cd} N/mm²	F.A.L. kN
1.000	26.810	218.871	2891.526
1.200	32.172	213.828	2824.904
1.400	37.534	208.466	2754.067
1.600	42.895	202.525	2675.580
1.800	48.257	196.091	2590.576
2.000	53.619	189.295	2500.790
2.200	58.981	182.324	2408.702
2.400	64.343	174.485	2305.138
2.600	69.705	166.442	2198.883
2.800	75.067	157.893	2085.934
3.000	80.429	149.314	1972.595
3.200	85.791	140.735	1859.256
3.400	91.153	132.155	1745.917
3.600	96.515	123.576	1632.578
3.800	101.877	115.373	1524.197
4.000	107.239	107.866	1425.026
4.200	112.601	100.801	1331.694
4.400	117.962	94.206	1244.565
4.600	123.324	88.143	1164.463
4.800	128.686	82.406	1088.667
5.000	134.048	77.276	1020.894
5.200	139.410	72.343	955.724
5.400	144.772	68.078	899.380
5.600	150.134	63.910	844.322
5.800	155.496	60.318	796.862
6.000	160.858	56.802	750.421
6.200	166.220	53.692	709.336
6.400	171.582	50.709	669.922
6.600	176.944	48.028	634.504
6.800	182.306	45.509	601.217
7.000	187.668	43.203	570.758
7.200	193.029	41.079	542.699
7.400	198.391	39.095	516.489
7.600	203.753	37.261	492.263
7.800	209.115	35.492	468.887
8.000	214.477	33.902	447.877
8.200	219.839	32.347	427.334
8.400	225.201	31.000	409.540
8.600	230.563	29.671	391.979
8.800	235.925	28.437	375.687
9.000	241.287	27.243	359.904
9.200	246.649	26.170	345.737
9.325	250.000	25.500	336.882

able XII (35) Axial Compression Load Capacity of I-Section

ISMB600 @122.6kg/m

Section	Area mm^2	$r_{min} = r_{yy}$ mm	Buckling Class y-y
ISMB600 @122.6kg/m	15621.242	41.2	b
Eff Length m	Sl Ratio	D.C.S. f_{cd} N/mm^2	F.A.L. kN
1.000	24.272	221.155	3454.721
1.200	29.126	216.786	3386.473
1.400	33.981	212.019	3312.007
1.600	38.835	207.165	3236.175
1.800	43.689	201.573	3148.818
2.000	48.544	195.748	3057.820
2.200	53.398	189.583	2961.515
2.400	58.252	183.272	2862.934
2.600	63.107	176.340	2754.647
2.800	67.961	169.058	2640.900
3.000	72.816	161.495	2522.755
3.200	77.670	153.728	2401.425
3.400	82.524	145.961	2280.095
3.600	87.379	138.194	2158.765
3.800	92.233	130.427	2037.435
4.000	97.087	122.660	1916.105
4.200	101.942	115.282	1800.841
4.400	106.796	108.485	1694.677
4.600	111.650	101.970	1592.897
4.800	116.505	95.999	1499.624
5.000	121.359	90.246	1409.749
5.200	126.214	85.051	1328.609
5.400	131.068	80.017	1249.972
5.600	135.922	75.551	1180.208
5.800	140.777	71.194	1112.141
6.000	145.631	67.408	1052.993
6.200	150.485	63.675	994.679
6.400	155.340	60.422	943.872
6.600	160.194	57.187	893.338
6.800	165.049	54.372	849.356
7.000	169.903	51.556	805.374
7.200	174.757	49.121	767.337
7.400	179.612	46.694	729.421
7.600	184.466	44.580	696.389
7.800	189.320	42.492	663.781
8.000	194.175	40.655	635.087
8.200	199.029	38.859	607.029
8.400	203.883	37.218	581.398
8.600	208.738	35.617	556.374
8.800	213.592	34.158	533.594
9.000	218.447	32.750	511.603

Contd...

9.200	223.301	31.475	491.675
9.400	228.155	30.261	472.717
9.600	233.010	29.108	454.699
9.800	237.864	27.991	437.258
10.000	242.718	26.956	421.091
10.200	247.573	25.985	405.925
10.300	250.000	25.500	398.342

Table XII (36) Axial Compression Load Capacity of I-Section

ISWB150 @17.0kg/m

Section	*Area mm²*	$r_{min} = r_{yy}$ *mm*	*Buckling Class y-y*
ISWB150 @17.0kg/m	*2166.534*	*20.9*	*b*
Eff Length m	*Sl Ratio*	*D.C.S.* f_{cd} *N/mm²*	*F.A.L. kN*
1.000	47.847	196.584	425.905
1.200	57.416	184.359	399.420
1.400	66.986	170.522	369.441
1.600	76.555	155.512	336.922
1.800	86.124	140.201	303.750
2.000	95.694	124.890	270.578
2.200	105.263	110.632	239.687
2.400	114.833	98.056	212.442
2.600	124.402	86.990	188.467
2.800	133.971	77.346	167.574
3.000	143.541	69.038	149.574
3.200	153.110	61.916	134.144
3.400	162.679	55.746	120.775
3.600	172.249	50.376	109.140
3.800	181.818	45.718	99.050
4.000	191.388	41.687	90.315
4.200	200.957	38.184	82.727
4.400	210.526	35.047	75.931
4.600	220.096	32.276	69.927
4.800	229.665	29.884	64.744
5.000	239.234	27.676	59.961
5.200	248.804	25.739	55.765
5.225	250.000	25.500	55.247

Table XII (37) Axial Compression Load Capacity of I-Section

ISWB175 @22.1kg/m

Section	Area mm²	$r_{min} = r_{yy}$ mm	Buckling Class y-y
ISWB175 @22.1kg/m	2811.294	25.9	b
Eff Length m	Sl Ratio	D.C.S. f_{cd} N/mm²	F.A.L. kN
1.000	38.610	207.390	583.034
1.200	46.332	198.402	557.765
1.400	54.054	188.730	530.575
1.600	61.776	178.336	501.355
1.800	69.498	166.753	468.791
2.000	77.220	154.448	434.198
2.200	84.942	142.093	399.464
2.400	92.664	129.737	364.730
2.600	100.386	117.459	330.213
2.800	108.108	106.649	299.821
3.000	115.830	96.829	272.215
3.200	123.552	87.899	247.111
3.400	131.274	79.828	224.419
3.600	138.996	72.724	204.447
3.800	146.718	66.560	187.119
4.000	154.440	61.025	171.560
4.200	162.162	56.046	157.562
4.400	169.884	51.567	144.971
4.600	177.606	47.697	134.090
4.800	185.328	44.209	124.284
5.000	193.050	41.071	115.464
5.200	200.772	38.245	107.518
5.400	208.494	35.697	100.355
5.600	216.216	33.397	93.890
5.800	223.938	31.315	88.037
6.000	231.660	29.418	82.703
6.200	239.382	27.642	77.710
6.400	247.104	26.079	73.316
6.475	250.000	25.500	71.688

Table XII (38) Axial Compression Load Capacity of I-Section

ISWB200 @28.8kg/m

Section	Area mm²	$r_{min} = r_{yy}$ mm	Buckling Class y-y
ISWB200 @28.8kg/m	3670.870	29.9	b
Eff Length m	Sl Ratio	D.C.S. f_{cd} N/mm²	F.A.L. kN
1.000	33.445	212.555	780.262
1.200	40.134	205.839	755.610
1.400	46.823	197.813	726.145

Contd...

1.600	53.512	189.435	695.390
1.800	60.201	180.699	663.323
2.000	66.890	170.666	626.491
2.200	73.579	160.274	588.346
2.400	80.268	149.572	549.059
2.600	86.957	138.870	509.772
2.800	93.645	128.167	470.485
3.000	100.334	117.532	431.444
3.200	107.023	108.167	397.068
3.400	113.712	99.434	365.008
3.600	120.401	91.271	335.042
3.800	127.090	84.113	308.769
4.000	133.779	77.523	284.577
4.200	140.468	71.435	262.228
4.400	147.157	66.217	243.075
4.600	153.846	61.423	225.476
4.800	160.535	56.990	209.202
5.000	167.224	53.110	194.960
5.200	173.913	49.543	181.868
5.400	180.602	46.241	169.745
5.600	187.291	43.365	159.187
5.800	193.980	40.727	149.505
6.000	200.669	38.279	140.518
6.200	207.358	36.072	132.415
6.400	214.047	34.026	124.907
6.600	220.736	32.116	117.894
6.800	227.425	30.444	111.755
7.000	234.114	28.854	105.919
7.200	240.803	27.339	100.360
7.400	247.492	26.002	95.449
7.475	250.000	25.500	93.607

Table XII (39) Axial Compression Load Capacity of I-Section

ISWB225 @33.9kg/m

Section	*Area mm²*	$r_{min} = r_{yy}$ *mm*	*Buckling Class y-y*
ISWB225 @33.9kg/m	*4323.950*	*32.2*	*b*
Eff Length m	*Sl Ratio*	*D.C.S.* f_{cd} *N/mm²*	*F.A.L. kN*
1.000	31.056	214.944	929.408
1.200	37.267	208.733	902.551
1.400	43.478	201.826	872.686
1.600	49.689	194.373	840.458
1.800	55.901	186.329	805.678
2.000	62.112	177.832	768.938
2.200	68.323	168.516	728.653

Contd...

2.400	74.534	158.745	686.407
2.600	80.745	148.807	643.436
2.800	86.957	138.870	600.465
3.000	93.168	128.932	557.494
3.200	99.379	118.994	514.523
3.400	105.590	110.174	476.386
3.600	111.801	101.784	440.111
3.800	118.012	94.145	407.077
4.000	124.224	87.181	376.965
4.200	130.435	80.600	348.510
4.400	136.646	74.886	323.802
4.600	142.857	69.571	300.823
4.800	149.068	64.727	279.875
5.000	155.280	60.463	261.438
5.200	161.491	56.435	244.024
5.400	167.702	52.833	228.447
5.600	173.913	49.543	214.224
5.800	180.124	46.447	200.833
6.000	186.335	43.776	189.284
6.200	192.547	41.258	178.397
6.400	198.758	38.960	168.459
6.600	204.969	36.860	159.382
6.800	211.180	34.858	150.723
7.000	217.391	33.057	142.935
7.200	223.602	31.399	135.769
7.400	229.814	29.847	129.055
7.600	236.025	28.414	122.862
7.800	242.236	27.053	116.975
8.000	248.447	25.811	111.604
8.050	250.000	25.500	110.261

able XII (40) Axial Compression Load Capacity of I-Section

ISWB250 @40.9kg/m

Section	*Area mm²*	*$r_{min} = r_{yy}$ mm*	*Buckling Class y-y*
ISWB250 @40.9kg/m	*5204.610*	*40.6*	*b*
Eff Length m	*Sl Ratio*	*D.C.S. f_{cd} N/mm²*	*F.A.L. kN*
1.000	24.631	220.833	1149.347
1.200	29.557	216.399	1126.272
1.400	34.483	211.517	1100.865
1.600	39.409	206.591	1075.226
1.800	44.335	200.798	1045.075
2.000	49.261	194.887	1014.309
2.200	54.187	188.557	981.364
2.400	59.113	182.153	948.034
2.600	64.039	174.941	910.499
2.800	68.966	167.552	872.041

Contd...

3.000	73.892	159.773	831.558
3.200	78.818	151.892	790.537
3.400	83.744	144.010	749.515
3.600	88.670	136.128	708.494
3.800	93.596	128.246	667.472
4.000	98.522	120.365	626.450
4.200	103.448	113.172	589.018
4.400	108.374	106.276	553.124
4.600	113.300	99.940	520.151
4.800	118.227	93.881	488.615
5.000	123.153	88.327	459.705
5.200	128.079	83.056	432.272
5.400	133.005	78.235	407.185
5.600	137.931	73.703	383.598
5.800	142.857	69.571	362.092
6.000	147.783	65.729	342.094
6.200	152.709	62.185	323.647
6.400	157.635	58.884	306.469
6.600	162.562	55.814	290.492
6.800	167.488	52.957	275.621
7.000	172.414	50.293	261.756
7.200	177.340	47.830	248.937
7.400	182.266	45.526	236.943
7.600	187.192	43.407	225.919
7.800	192.118	41.416	215.555
8.000	197.044	39.594	206.069
8.200	201.970	37.850	196.993
8.400	206.897	36.224	188.533
8.600	211.823	34.671	180.451
8.800	216.749	33.243	173.016
9.000	221.675	31.881	165.930
9.200	226.601	30.650	159.520
9.400	231.527	29.449	153.269
9.600	236.453	28.316	147.372
9.800	241.379	27.224	141.691
10.000	246.305	26.239	136.563
10.150	250.000	25.500	132.718

Table XII (41) Axial Compression Load Capacity of I-Section

ISWB300 @48.1kg/m

Section	*Area mm²*	*$r_{min} = r_{yy}$ mm*	*Buckling Class y-y*
ISWB300 @48.1kg/m	*6132.754*	*40.2*	*b*
Eff Length m	*Sl Ratio*	*D.C.S. f_{cd} N/mm²*	*F.A.L. kN*
1.000	24.876	220.612	1352.959
1.200	29.851	216.134	1325.499
1.400	34.826	211.174	1295.079

Contd...

1.600	39.801	206.199	1264.568
1.800	44.776	200.269	1228.198
2.000	49.751	194.299	1191.585
2.200	54.726	187.856	1152.073
2.400	59.701	181.388	1112.408
2.600	64.677	173.985	1067.008
2.800	69.652	166.522	1021.241
3.000	74.627	158.597	972.636
3.200	79.602	150.637	923.819
3.400	84.577	142.677	875.001
3.600	89.552	134.716	826.183
3.800	94.527	126.756	777.365
4.000	99.502	118.796	728.547
4.200	104.478	111.731	685.221
4.400	109.453	104.766	642.505
4.600	114.428	98.554	604.406
4.800	119.403	92.434	566.877
5.000	124.378	87.015	533.644
5.200	129.353	81.692	500.997
5.400	134.328	77.018	472.332
5.600	139.303	72.441	444.262
5.800	144.279	68.463	419.865
6.000	149.254	64.582	396.066
6.200	154.229	61.167	375.120
6.400	159.204	57.833	354.678
6.600	164.179	54.876	336.542
6.800	169.154	51.991	318.845
7.000	174.129	49.435	303.175
7.200	179.104	46.948	287.919
7.400	184.080	44.746	274.415
7.600	189.055	42.606	261.295
7.800	194.030	40.709	249.658
8.000	199.005	38.868	238.369
8.200	203.980	37.187	228.056
8.400	208.955	35.545	217.987
8.600	213.930	34.060	208.883
8.800	218.905	32.617	200.035
9.000	223.881	31.330	192.138
9.200	228.856	30.086	184.510
9.400	233.831	28.919	177.353
9.600	238.806	27.775	170.335
9.800	243.781	26.744	164.013
10.000	248.756	25.749	157.911
10.050	250.000	25.500	156.385

Table XII (42) Axial Compression Load Capacity of I-Section

ISWB350 @56.9kg/m

Section	Area mm^2	$r_{min} = r_{yy}$ mm	Buckling Class y-y
ISWB350 @56.9kg/m	7249.902	40.3	b
Eff Length m	Sl Ratio	D.C.S. f_{cd} N/mm²	F.A.L. kN
1.000	24.814	220.667	1599.818
1.200	29.777	216.201	1567.436
1.400	34.739	211.261	1531.618
1.600	39.702	206.298	1495.639
1.800	44.665	200.402	1452.895
2.000	49.628	194.447	1409.719
2.200	54.591	188.032	1363.215
2.400	59.553	181.581	1316.442
2.600	64.516	174.226	1263.120
2.800	69.479	166.782	1209.151
3.000	74.442	158.893	1151.961
3.200	79.404	150.953	1094.393
3.400	84.367	143.012	1036.826
3.600	89.330	135.072	979.258
3.800	94.293	127.132	921.691
4.000	99.256	119.191	864.124
4.200	104.218	112.094	812.673
4.400	109.181	105.146	762.301
4.600	114.144	98.903	717.037
4.800	119.107	92.799	672.782
5.000	124.069	87.346	633.247
5.200	129.032	82.035	594.749
5.400	133.995	77.325	560.596
5.600	138.958	72.759	527.494
5.800	143.921	68.742	498.372
6.000	148.883	64.871	470.308
6.200	153.846	61.423	445.311
6.400	158.809	58.098	421.205
6.600	163.772	55.112	399.560
6.800	168.734	52.234	378.691
7.000	173.697	49.651	359.968
7.200	178.660	47.170	341.978
7.400	183.623	44.942	325.826
7.600	188.586	42.808	310.355
7.800	193.548	40.887	296.427
8.000	198.511	39.051	283.115
8.200	203.474	37.354	270.810
8.400	208.437	35.716	258.937
8.600	213.400	34.214	248.049
8.800	218.362	32.775	237.615
9.000	223.325	31.469	228.145

Contd...

9.200	228.288	30.228	219.150
9.400	233.251	29.052	210.627
9.600	238.213	27.911	202.351
9.800	243.176	26.865	194.767
10.000	248.139	25.872	187.571
10.075	250.000	25.500	184.873

able XII (43) Axial Compression Load Capacity of I-Section

ISWB400 @66.7kg/m

Section	*Area mm²*	$r_{min} = r_{yy}$ *mm*	*Buckling Class y-y*
ISWB400 @66.7kg/m	*8501.254*	*40.4*	*b*
Eff Length m	*Sl Ratio*	*D.C.S.* f_{cd} *N/mm²*	*F.A.L. kN*
1.000	24.752	220.723	1876.420
1.200	29.703	216.267	1838.544
1.400	34.653	211.347	1796.711
1.600	39.604	206.396	1754.625
1.800	44.554	200.535	1704.796
2.000	49.505	194.594	1654.294
2.200	54.455	188.208	1600.003
2.400	59.406	181.772	1545.292
2.600	64.356	174.465	1483.174
2.800	69.307	167.040	1420.046
3.000	74.257	159.188	1353.299
3.200	79.208	151.267	1285.962
3.400	84.158	143.347	1218.625
3.600	89.109	135.426	1151.289
3.800	94.059	127.505	1083.952
4.000	99.010	119.584	1016.615
4.200	103.960	112.455	956.012
4.400	108.911	105.525	897.093
4.600	113.861	99.250	843.754
4.800	118.812	93.161	791.989
5.000	123.762	87.674	745.341
5.200	128.713	82.377	700.310
5.400	133.663	77.630	659.950
5.600	138.614	73.075	621.231
5.800	143.564	69.020	586.755
6.000	148.515	65.158	553.928
6.200	153.465	61.678	524.342
6.400	158.416	58.361	496.145
6.600	163.366	55.348	470.523
6.800	168.317	52.476	446.114
7.000	173.267	49.866	423.926
7.200	178.218	47.391	402.884
7.400	183.168	45.138	383.726
7.600	188.119	43.009	365.630

Contd...

7.800	193.069	41.064	349.099
8.000	198.020	39.233	333.527
8.200	202.970	37.520	318.965
8.400	207.921	35.886	305.077
8.600	212.871	34.367	292.165
8.800	217.822	32.932	279.961
9.000	222.772	31.607	268.699
9.200	227.723	30.369	258.177
9.400	232.673	29.185	248.110
9.600	237.624	28.047	238.431
9.800	242.574	26.985	229.408
10.000	247.525	25.995	220.991
10.100	250.000	25.500	216.782

Table XII (44) Axial Compression Load Capacity of I-Section

ISWB450 @79.4kg/m

Section	*Area mm²*	*$r_{min} = r_{yy}$ mm*	*Buckling Class y-y*
ISWB450 @79.4kg/m	*10115.051*	*41.1*	*b*
Eff Length m	*Sl Ratio*	*D.C.S. f_{cd} N/mm²*	*F.A.L. kN*
1.000	24.331	221.102	2236.460
1.200	29.197	216.723	2192.160
1.400	34.063	211.937	2143.751
1.600	38.929	207.071	2094.529
1.800	43.796	201.445	2037.629
2.000	48.662	195.606	1978.563
2.200	53.528	189.414	1915.928
2.400	58.394	183.088	1851.940
2.600	63.260	176.109	1781.356
2.800	68.127	168.810	1707.524
3.000	72.993	161.212	1630.664
3.200	77.859	153.426	1551.910
3.400	82.725	145.640	1473.155
3.600	87.591	137.854	1394.400
3.800	92.457	130.068	1315.646
4.000	97.324	122.282	1236.891
4.200	102.190	114.934	1162.566
4.400	107.056	108.122	1093.656
4.600	111.922	101.636	1028.051
4.800	116.788	95.650	967.508
5.000	121.655	89.930	909.643
5.200	126.521	84.723	856.976
5.400	131.387	79.724	806.413
5.600	136.253	75.247	761.129
5.800	141.119	70.927	717.430
6.000	145.985	67.131	679.037
6.200	150.852	63.429	641.592

Contd...

6.400	155.718	60.169	608.614
6.600	160.584	56.961	576.167
6.800	165.450	54.139	547.618
7.000	170.316	51.342	519.325
7.200	175.182	48.909	494.715
7.400	180.049	46.479	470.138
7.600	184.915	44.387	448.973
7.800	189.781	42.294	427.808
8.000	194.647	40.481	409.463
8.200	199.513	38.680	391.251
8.400	204.380	37.055	374.811
8.600	209.246	35.449	358.567
8.800	214.112	34.008	343.988
9.000	218.978	32.596	329.714
9.200	223.844	31.339	316.995
9.400	228.710	30.122	304.689
9.600	233.577	28.977	293.108
9.800	238.443	27.858	281.787
10.000	243.309	26.838	271.470
10.200	248.175	25.865	261.625
10.275	250.000	25.500	257.934

able XII (45) Axial Compression Load Capacity of I-Section

ISWB500 @95.2kg/m

Section	*Area mm²*	$r_{min} = r_{yy}$ *mm*	*Buckling Class y-y*
ISWB500 @95.2kg/m	*12121.912*	*49.6*	*b*
Eff Length m	*Sl Ratio*	*D.C.S.* f_{cd} *N/mm²*	*F.A.L. kN*
1.000	20.161	224.855	2725.671
1.200	24.194	221.226	2681.680
1.400	28.226	217.597	2637.689
1.600	32.258	213.742	2590.961
1.800	36.290	209.710	2542.082
2.000	40.323	205.613	2492.421
2.200	44.355	200.774	2433.767
2.400	48.387	195.935	2375.113
2.600	52.419	190.855	2313.526
2.800	56.452	185.613	2249.983
3.000	60.484	180.274	2185.268
3.200	64.516	174.226	2111.950
3.400	68.548	168.177	2038.632
3.600	72.581	161.871	1962.186
3.800	76.613	155.419	1883.980
4.000	80.645	148.968	1805.774
4.200	84.677	142.516	1727.568
4.400	88.710	136.065	1649.362

Contd...

4.600	92.742	129.613	1571.156
4.800	96.774	123.161	1492.950
5.000	100.806	116.871	1416.700
5.200	104.839	111.226	1348.269
5.400	108.871	105.581	1279.839
5.600	112.903	100.429	1217.392
5.800	116.935	95.469	1157.271
6.000	120.968	90.665	1099.027
6.200	125.000	86.350	1046.727
6.400	129.032	82.035	994.427
6.600	133.065	78.181	947.699
6.800	137.097	74.471	902.731
7.000	141.129	70.919	859.678
7.200	145.161	67.774	821.553
7.400	149.194	64.629	783.427
7.600	153.226	61.839	749.603
7.800	157.258	59.137	716.855
8.000	161.290	56.552	685.514
8.200	165.323	54.213	657.164
8.400	169.355	51.874	628.814
8.600	173.387	49.806	603.749
8.800	177.419	47.790	579.310
9.000	181.452	45.876	556.102
9.200	185.484	44.142	535.085
9.400	189.516	42.408	514.067
9.600	193.548	40.887	495.630
9.800	197.581	39.395	477.545
10.000	201.613	37.968	460.242
10.200	205.645	36.637	444.112
10.400	209.677	35.306	427.982
10.600	213.710	34.124	413.650
10.800	217.742	32.955	399.476
11.000	221.774	31.856	386.161
11.200	225.806	30.848	373.941
11.400	229.839	29.840	361.722
11.600	233.871	28.910	350.441
11.800	237.903	27.982	339.198
12.000	241.935	27.113	328.660
12.200	245.968	26.306	318.884
12.400	250.000	25.500	309.109

Table XII (46) Axial Compression Load Capacity of I-Section

ISWB550@112.5kg/m

Section	Area mm²	$r_{min} = r_{yy}$ mm	Buckling Class y-y
ISWB550@112.5kg/m	14333.937	51.1	b
Eff Length m	Sl Ratio	D.C.S. f_{cd} N/mm²	F.A.L. kN
1.000	19.569	225.086	3226.370
1.200	23.483	221.865	3180.198
1.400	27.397	218.342	3129.707
1.600	31.311	214.689	3077.336
1.800	35.225	210.775	3021.235
2.000	39.139	206.861	2965.133
2.200	43.053	202.337	2900.280
2.400	46.967	197.640	2832.958
2.600	50.881	192.855	2764.374
2.800	54.795	187.767	2691.442
3.000	58.708	182.679	2618.510
3.200	62.622	177.067	2538.061
3.400	66.536	171.196	2453.908
3.600	70.450	165.280	2369.111
3.800	74.364	159.018	2279.348
4.000	78.278	152.755	2189.586
4.200	82.192	146.493	2099.824
4.400	86.106	140.231	2010.061
4.600	90.020	133.969	1920.299
4.800	93.933	127.706	1830.536
5.000	97.847	121.444	1740.774
5.200	101.761	115.534	1656.061
5.400	105.675	110.055	1577.518
5.600	109.589	104.575	1498.976
5.800	113.503	99.691	1428.970
6.000	117.417	94.877	1359.965
6.200	121.331	90.276	1294.012
6.400	125.245	86.088	1233.984
6.600	129.159	81.900	1173.955
6.800	133.072	78.173	1120.532
7.000	136.986	74.573	1068.919
7.200	140.900	71.098	1019.112
7.400	144.814	68.045	975.353
7.600	148.728	64.992	931.594
7.800	152.642	62.230	892.000
8.000	156.556	59.608	854.412
8.200	160.470	57.028	817.430
8.400	164.384	54.758	784.891

8.600	168.297	52.487	752.352
8.800	172.211	50.394	722.349
9.000	176.125	48.437	694.298
9.200	180.039	46.483	666.287
9.400	183.953	44.800	642.163
9.600	187.867	43.117	618.040
9.800	191.781	41.541	595.447
10.000	195.695	40.093	574.690
10.200	199.609	38.645	553.932
10.400	203.523	37.338	535.194
10.600	207.436	36.046	516.681
10.800	211.350	34.808	498.942
11.000	215.264	33.673	482.672
11.200	219.178	32.538	466.403
11.400	223.092	31.527	451.906
11.600	227.006	30.549	437.881
11.800	230.920	29.588	424.119
12.000	234.834	28.688	411.216
12.200	238.748	27.788	398.312
12.400	242.661	26.968	386.553
12.600	246.575	26.185	375.333
12.775	250.000	25.500	365.515

Table XII (47) Axial Compression Load Capacity of I-Section

ISWB600 @133.7kg/m

Section	*Area mm²*	*$r_{min} = r_{yy}$ mm*	*Buckling Class y-y*
ISWB600 @133.7kg/m	*17037.986*	*52.5*	*b*
Eff Length m	*Sl Ratio*	*D.C.S. f_{cd} N/mm²*	*F.A.L. kN*
1.000	19.048	225.190	3836.792
1.200	22.857	222.429	3789.735
1.400	26.667	219.000	3731.319
1.600	30.476	215.524	3672.092
1.800	34.286	211.714	3607.185
2.000	38.095	207.905	3542.278
2.200	41.905	203.714	3470.881
2.400	45.714	199.143	3392.993
2.600	49.524	194.571	3315.105
2.800	53.333	189.667	3231.538
3.000	57.143	184.714	3147.159
3.200	60.952	179.571	3059.536
3.400	64.762	173.857	2962.176
3.600	68.571	168.143	2864.816
3.800	72.381	162.190	2763.399

Contd...

4.000	76.190	156.095	2659.548
4.200	80.000	150.000	2555.698
4.400	83.810	143.905	2451.847
4.600	87.619	137.810	2347.997
4.800	91.429	131.714	2244.146
5.000	95.238	125.619	2140.296
5.200	99.048	119.524	2036.445
5.400	102.857	114.000	1942.330
5.600	106.667	108.667	1851.461
5.800	110.476	103.414	1761.971
6.000	114.286	98.729	1682.136
6.200	118.095	94.043	1602.301
6.400	121.905	89.662	1527.658
6.600	125.714	85.586	1458.208
6.800	129.524	81.510	1388.758
7.000	133.333	77.933	1327.827
7.200	137.143	74.429	1268.113
7.400	140.952	71.057	1210.671
7.600	144.762	68.086	1160.043
7.800	148.571	65.114	1109.416
8.000	152.381	62.405	1063.251
8.200	156.190	59.852	1019.764
8.400	160.000	57.300	976.277
8.600	163.810	55.090	938.631
8.800	167.619	52.881	900.985
9.000	171.429	50.786	865.286
9.200	175.238	48.881	832.833
9.400	179.048	46.976	800.380
9.600	182.857	45.271	771.334
9.800	186.667	43.633	743.424
10.000	190.476	42.024	716.001
10.200	194.286	40.614	691.986
10.400	198.095	39.205	667.970
10.600	201.905	37.871	645.253
10.800	205.714	36.614	623.834
11.000	209.524	35.357	602.415
11.200	213.333	34.233	583.267
11.400	217.143	33.129	564.444
11.600	220.952	32.062	546.270
11.800	224.762	31.110	530.044
12.000	228.571	30.157	513.817
12.200	232.381	29.252	498.402
12.400	236.190	28.376	483.473

Contd...

12.600	240.000	27.500	468.545
12.800	243.810	26.738	455.563
13.000	247.619	25.976	442.582
13.125	250.000	25.500	434.469

Table XII (48) Axial Compression Load Capacity of I-Section

ISWB600 @145.1kg/m

Section	*Area mm²*	$r_{min} = r_{yy}$ *mm*	*Buckling Class y-y*
ISWB600 @145.1kg/m	*18514.040*	*53.5*	*b*
Eff Length m	*Sl Ratio*	*D.C.S.* f_{cd} *N/mm²*	*F.A.L. kN*
1.000	18.692	225.262	4170.504
1.200	22.430	222.813	4125.170
1.400	26.168	219.449	4062.880
1.600	29.907	216.084	4000.590
1.800	33.645	212.355	3931.551
2.000	37.383	208.617	3862.340
2.200	41.121	204.654	3788.976
2.400	44.860	200.168	3705.922
2.600	48.598	195.682	3622.869
2.800	52.336	190.963	3535.489
3.000	56.075	186.103	3445.515
3.200	59.813	181.243	3355.540
3.400	63.551	175.673	3252.415
3.600	67.290	170.065	3148.598
3.800	71.028	164.355	3042.878
4.000	74.766	158.374	2932.139
4.200	78.505	152.393	2821.401
4.400	82.243	146.411	2710.663
4.600	85.981	140.430	2599.925
4.800	89.720	134.449	2489.187
5.000	93.458	128.467	2378.448
5.200	97.196	122.486	2267.710
5.400	100.935	116.692	2160.433
5.600	104.673	111.458	2063.537
5.800	108.411	106.224	1966.641
6.000	112.150	101.356	1876.510
6.200	115.888	96.758	1791.380
6.400	119.626	92.160	1706.250
6.600	123.364	88.100	1631.087
6.800	127.103	84.100	1557.031
7.000	130.841	80.226	1485.310
7.200	134.579	76.787	1421.636
7.400	138.318	73.348	1357.962

Contd...

7.600	142.056	70.196	1299.616
7.800	145.794	67.280	1245.631
8.000	149.533	64.364	1191.647
8.200	153.271	61.808	1144.323
8.400	157.009	59.304	1097.952
8.600	160.748	56.866	1052.826
8.800	164.486	54.698	1012.683
9.000	168.224	52.530	972.541
9.200	171.963	50.519	935.305
9.400	175.701	48.650	900.699
9.600	179.439	46.780	866.094
9.800	183.178	45.134	835.606
10.000	186.916	43.526	805.845
10.200	190.654	41.958	776.811
10.400	194.393	40.575	751.203
10.600	198.131	39.192	725.595
10.800	201.869	37.883	701.371
11.000	205.607	36.650	678.531
11.200	209.346	35.416	655.691
11.400	213.084	34.306	635.135
11.600	216.822	33.221	615.064
11.800	220.561	32.160	595.408
12.000	224.299	31.225	578.105
12.200	228.037	30.291	560.802
12.400	231.776	29.392	544.157
12.600	235.514	28.532	528.238
12.800	239.252	27.672	512.320
13.000	242.991	26.902	498.062
13.200	246.729	26.154	484.220
13.375	250.000	25.500	472.108

able XII (49) Axial Compression Load Capacity of I-Section

ISHB150 @27.1kg/m

Section	*Area mm²*	*$r_{min} = r_{yy}$ mm*	*Buckling Class y-y*
ISHB150 @27.1kg/m	*3447.749*	*35.4*	*c*
Eff Length m	*Sl Ratio*	*D.C.S. f_{cd} N/mm²*	*F.A.L. kN*
1.000	28.249	213.277	735.325
1.200	33.898	205.932	710.002
1.400	39.548	198.588	684.680
1.600	45.198	190.203	655.773
1.800	50.847	181.729	626.555
2.000	56.497	173.254	597.337
2.200	62.147	164.565	567.379
2.400	67.797	155.525	536.213

Contd...

2.600	73.446	146.486	505.046
2.800	79.096	137.446	473.880
3.000	84.746	128.881	444.351
3.200	90.395	120.446	415.269
3.400	96.045	112.537	387.998
3.600	101.695	104.898	361.663
3.800	107.345	97.893	337.509
4.000	112.994	91.336	314.904
4.200	118.644	85.178	293.672
4.400	124.294	79.664	274.661
4.600	129.944	74.353	256.351
4.800	135.593	69.769	240.548
5.000	141.243	65.330	225.241
5.200	146.893	61.375	211.606
5.400	152.542	57.700	198.935
5.600	158.192	54.367	187.443
5.800	163.842	51.302	176.877
6.000	169.492	48.364	166.748
6.200	175.141	45.786	157.860
6.400	180.791	43.292	149.258
6.600	186.441	41.088	141.662
6.800	192.090	38.989	134.425
7.000	197.740	37.068	127.802
7.200	203.390	35.283	121.647
7.400	209.040	33.588	115.803
7.600	214.689	31.893	109.960
7.800	220.339	30.522	105.232
8.000	225.989	29.223	100.752
8.200	231.638	27.956	96.385
8.400	237.288	26.769	92.294
8.600	242.938	25.642	88.407
8.800	248.588	24.568	84.706
8.850	250.000	24.300	83.780

Table XII (50) Axial Compression Load Capacity of I-Section

ISHB150 @30.7kg/m

Section	*Area mm²*	$r_{min} = r_{yy}$ *mm*	*Buckling Class y-y*
ISHB150 @30.7kg/m	*3897.749*	*34.4*	*c*
Eff Length m	*Sl Ratio*	*D.C.S.* f_{cd} *N/mm²*	*F.A.L. kN*
1.000	29.070	212.209	827.138
1.200	34.884	204.651	797.679
1.400	40.698	196.953	767.675
1.600	46.512	188.233	733.683
1.800	52.326	179.512	699.691
2.000	58.140	170.791	665.699
2.200	63.953	161.674	630.166

Contd...

2.400	69.767	152.372	593.908
2.600	75.581	143.070	557.650
2.800	81.395	133.907	521.936
3.000	87.209	125.186	487.944
3.200	93.023	116.767	455.130
3.400	98.837	108.628	423.404
3.600	104.651	101.233	394.579
3.800	110.465	94.093	366.751
4.000	116.279	87.756	342.050
4.200	122.093	81.733	318.573
4.400	127.907	76.267	297.271
4.600	133.721	71.286	277.855
4.800	139.535	66.577	259.499
5.000	145.349	62.456	243.437
5.200	151.163	58.514	228.073
5.400	156.977	55.084	214.702
5.600	162.791	51.849	202.094
5.800	168.605	48.826	190.310
6.000	174.419	46.112	179.732
6.200	180.233	43.509	169.588
6.400	186.047	41.242	160.750
6.600	191.860	39.067	152.275
6.800	197.674	37.091	144.570
7.000	203.488	35.253	137.409
7.200	209.302	33.509	130.611
7.400	215.116	31.919	124.411
7.600	220.930	30.386	118.437
7.800	226.744	29.049	113.225
8.000	232.558	27.763	108.212
8.200	238.372	26.542	103.453
8.400	244.186	25.405	99.021
8.600	250.000	24.300	94.715

able XII (51) Axial Compression Load Capacity of I-Section

ISHB150 @34.6kg/m

Section	*Area mm²*	*$r_{min} = r_{yy}$ mm*	*Buckling Class y-y*
ISHB150 @34.6kg/m	*4407.749*	*33.5*	*c*
Eff Length m	*Sl Ratio*	*D.C.S. f_{cd} N/mm²*	*F.A.L. kN*
1.000	29.851	211.194	930.890
1.200	35.821	203.433	896.681
1.400	41.791	195.313	860.892
1.600	47.761	186.358	821.420
1.800	53.731	177.403	781.948
2.000	59.701	168.448	742.475
2.200	65.672	158.925	700.503
2.400	71.642	149.373	658.399

Contd...

2.600	77.612	139.821	616.295
2.800	83.582	130.627	575.770
3.000	89.552	121.672	536.298
3.200	95.522	113.269	499.260
3.400	101.493	105.149	463.471
3.600	107.463	97.746	430.841
3.800	113.433	90.858	400.480
4.000	119.403	84.351	371.797
4.200	125.373	78.649	346.666
4.400	131.343	73.212	322.700
4.600	137.313	68.376	301.385
4.800	143.284	63.901	281.662
5.000	149.254	59.722	263.241
5.200	155.224	56.118	247.354
5.400	161.194	52.679	232.196
5.600	167.164	49.575	218.512
5.800	173.134	46.690	205.796
6.000	179.104	44.003	193.954
6.200	185.075	41.621	183.454
6.400	191.045	39.345	173.422
6.600	197.015	37.315	164.475
6.800	202.985	35.404	156.054
7.000	208.955	33.613	148.160
7.200	214.925	31.970	140.916
7.400	220.896	30.394	133.969
7.600	226.866	29.021	127.917
7.800	232.836	27.704	122.114
8.000	238.806	26.451	116.588
8.200	244.776	25.293	111.483
8.375	250.000	24.300	107.108

Table XII (52) Axial Compression Load Capacity of I-Section

ISHB200 @ 37.3kg/m

Section	*Area mm²*	$r_{min} = r_{yy}$ *mm*	*Buckling Class y-y*
ISHB200 @ 37.3kg/m	*4754.432*	*45.1*	*c*
Eff Length m	*Sl Ratio*	*D.C.S.* f_{cd} *N/mm²*	*F.A.L. kN*
1.000	22.173	221.175	1051.562
1.200	26.608	215.410	1024.153
1.400	31.042	209.645	996.744
1.600	35.477	203.880	969.335
1.800	39.911	198.115	941.926
2.000	44.346	191.481	910.384
2.200	48.780	184.829	878.758
2.400	53.215	178.177	847.132
2.600	57.650	171.525	815.506

Contd...

2.800	62.084	164.665	782.889
3.000	66.519	157.570	749.155
3.200	70.953	150.475	715.421
3.400	75.388	143.379	681.686
3.600	79.823	136.284	647.952
3.800	84.257	129.614	616.242
4.000	88.692	122.962	584.616
4.200	93.126	116.623	554.476
4.400	97.561	110.415	524.959
4.600	101.996	104.525	496.959
4.800	106.430	99.027	470.815
5.000	110.865	93.657	445.288
5.200	115.299	88.824	422.306
5.400	119.734	83.990	399.325
5.600	124.169	79.782	379.316
5.800	128.603	75.613	359.497
6.000	133.038	71.839	341.556
6.200	137.472	68.247	324.478
6.400	141.907	64.865	308.397
6.600	146.341	61.761	293.638
6.800	150.776	58.742	279.285
7.000	155.211	56.126	266.846
7.200	159.645	53.509	254.406
7.400	164.080	51.178	243.325
7.600	168.514	48.873	232.361
7.800	172.949	46.773	222.379
8.000	177.384	44.777	212.891
8.200	181.818	42.891	203.922
8.400	186.253	41.161	195.699
8.600	190.687	39.466	187.640
8.800	195.122	37.959	180.471
9.000	199.557	36.451	173.303
9.200	203.991	35.103	166.893
9.400	208.426	33.772	160.568
9.600	212.860	32.528	154.651
9.800	217.295	31.330	148.958
10.000	221.729	30.202	143.594
10.200	226.164	29.182	138.745
10.400	230.599	28.174	133.953
10.600	235.033	27.243	129.525
10.800	239.468	26.312	125.097
11.000	243.902	25.459	121.041
11.200	248.337	24.616	117.035
11.275	250.000	24.300	115.533

Table XII (53) Axial Compression Load Capacity of I-Section

ISHB200 @40.0kg/m

Section	Area mm^2	$r_{min} = r_{yy}$ mm	Buckling Class y-y
ISHB200 @40.0kg/m	5094.432	44.2	c
Eff Length m	Sl Ratio	D.C.S. f_{cd} N/mm^2	F.A.L. kN
1.000	22.624	220.588	1123.772
1.200	27.149	214.706	1093.804
1.400	31.674	208.824	1063.837
1.600	36.199	202.941	1033.870
1.800	40.724	196.914	1003.165
2.000	45.249	190.127	968.587
2.200	49.774	183.339	934.010
2.400	54.299	176.552	899.432
2.600	58.824	169.765	864.855
2.800	63.348	162.643	828.571
3.000	67.873	155.403	791.689
3.200	72.398	148.163	754.806
3.400	76.923	140.923	717.923
3.600	81.448	133.828	681.778
3.800	85.973	127.041	647.200
4.000	90.498	120.303	612.876
4.200	95.023	113.968	580.604
4.400	99.548	107.633	548.331
4.600	104.072	101.950	519.378
4.800	108.597	96.339	490.794
5.000	113.122	91.197	464.596
5.200	117.647	86.265	439.470
5.400	122.172	81.658	416.003
5.600	126.697	77.405	394.334
5.800	131.222	73.310	373.475
6.000	135.747	69.645	354.803
6.200	140.271	66.010	336.283
6.400	144.796	62.843	320.147
6.600	149.321	59.675	304.011
6.800	153.846	56.931	290.030
7.000	158.371	54.261	276.429
7.200	162.896	51.794	263.862
7.400	167.421	49.441	251.875
7.600	171.946	47.224	240.582
7.800	176.471	45.188	230.208
8.000	180.995	43.212	220.139
8.200	185.520	41.447	211.149
8.400	190.045	39.685	202.171

Contd...

8.600	194.570	38.146	194.333
8.800	199.095	36.608	186.495
9.000	203.620	35.214	179.395
9.200	208.145	33.857	172.480
9.400	212.670	32.579	165.972
9.600	217.195	31.357	159.748
9.800	221.719	30.205	153.875
10.000	226.244	29.164	148.573
10.200	230.769	28.138	143.349
10.400	235.294	27.188	138.509
10.600	239.819	26.238	133.668
10.800	244.344	25.375	129.269
11.000	248.869	24.515	124.890
11.050	250.000	24.300	123.795

able XII (54) Axial Compression Load Capacity of I-Section

ISHB225 @43.1kg/m

Section	*Area mm²*	$r_{min} = r_{yy}$ *mm*	*Buckling Class y-y*
ISHB225 @43.1kg/m	*5493.807*	*49.6*	*c*
Eff Length m	*Sl Ratio*	*D.C.S.* f_{cd} *N/mm²*	*F.A.L. kN*
1.000	20.161	223.790	1229.461
1.200	24.194	218.548	1200.663
1.400	28.226	213.306	1171.864
1.600	32.258	208.065	1143.066
1.800	36.290	202.823	1114.268
2.000	40.323	197.516	1085.116
2.200	44.355	191.468	1051.887
2.400	48.387	185.419	1018.658
2.600	52.419	179.371	985.429
2.800	56.452	173.323	952.201
3.000	60.484	167.226	918.706
3.200	64.516	160.774	883.262
3.400	68.548	154.323	847.818
3.600	72.581	147.871	812.375
3.800	76.613	141.419	776.931
4.000	80.645	135.032	741.841
4.200	84.677	128.984	708.612
4.400	88.710	122.935	675.384
4.600	92.742	117.161	643.662
4.800	96.774	111.516	612.648
5.000	100.806	106.000	582.344
5.200	104.839	101.000	554.875
5.400	108.871	96.000	527.405

Contd...

5.600	112.903	91.435	502.329
5.800	116.935	87.040	478.183
6.000	120.968	82.790	454.834
6.200	125.000	79.000	434.011
6.400	129.032	75.210	413.187
6.600	133.065	71.818	394.553
6.800	137.097	68.552	376.609
7.000	141.129	65.410	359.348
7.200	145.161	62.587	343.841
7.400	149.194	59.765	328.335
7.600	153.226	57.297	314.777
7.800	157.258	54.918	301.707
8.000	161.290	52.629	289.134
8.200	165.323	50.532	277.614
8.400	169.355	48.435	266.095
8.600	173.387	46.576	255.878
8.800	177.419	44.761	245.910
9.000	181.452	43.034	236.420
9.200	185.484	41.461	227.780
9.400	189.516	39.889	219.141
9.600	193.548	38.494	211.476
9.800	197.581	37.123	203.944
10.000	201.613	35.816	196.767
10.200	205.645	34.606	190.121
10.400	209.677	33.397	183.475
10.600	213.710	32.298	177.441
10.800	217.742	31.210	171.460
11.000	221.774	30.192	165.869
11.200	225.806	29.265	160.774
11.400	229.839	28.337	155.679
11.600	233.871	27.487	151.009
11.800	237.903	26.640	146.357
12.000	241.935	25.832	141.917
12.200	245.968	25.066	137.708
12.400	250.000	24.300	133.500

able XII (55) Axial Compression Load Capacity of I-Section

ISHB225 @46.8kg/m

Section	Area mm²	$r_{min} = r_{yy}$ mm	Buckling Class y-y
ISHB225 @46.8kg/m	5966.307	48.4	c
Eff Length m	Sl Ratio	D.C.S. f_{cd} N/mm²	F.A.L. kN
1.000	20.661	223.140	1331.325
1.200	24.793	217.769	1299.274
1.400	28.926	212.397	1267.224
1.600	33.058	207.025	1235.173
1.800	37.190	201.653	1203.123
2.000	41.322	196.017	1169.495
2.200	45.455	189.818	1132.514
2.400	49.587	183.620	1095.532
2.600	53.719	177.421	1058.551
2.800	57.851	171.223	1021.570
3.000	61.983	164.826	983.405
3.200	66.116	158.215	943.959
3.400	70.248	151.603	904.512
3.600	74.380	144.992	865.065
3.800	78.512	138.380	825.619
4.000	82.645	132.033	787.750
4.200	86.777	125.835	750.769
4.400	90.909	119.727	714.330
4.600	95.041	113.942	679.814
4.800	99.174	108.157	645.298
5.000	103.306	102.901	613.938
5.200	107.438	97.777	583.367
5.400	111.570	92.888	554.201
5.600	115.702	88.384	527.328
5.800	119.835	83.880	500.455
6.000	123.967	79.971	477.132
6.200	128.099	76.087	453.957
6.400	132.231	72.493	432.513
6.600	136.364	69.145	412.543
6.800	140.496	65.853	392.899
7.000	144.628	62.960	375.641
7.200	148.760	60.068	358.383
7.400	152.893	57.493	343.023
7.600	157.025	55.055	328.477
7.800	161.157	52.698	314.415
8.000	165.289	50.550	301.594
8.200	169.421	48.401	288.774
8.400	173.554	46.501	277.438

Contd...

8.600	177.686	44.641	266.344
8.800	181.818	42.891	255.900
9.000	185.950	41.279	246.285
9.200	190.083	39.672	236.695
9.400	194.215	38.267	228.312
9.600	198.347	36.862	219.930
9.800	202.479	35.556	212.139
10.000	206.612	34.317	204.743
10.200	210.744	33.099	197.480
10.400	214.876	31.983	190.823
10.600	219.008	30.868	184.167
10.800	223.140	29.878	178.259
11.000	227.273	28.927	172.589
11.200	231.405	28.005	167.086
11.400	235.537	27.137	161.909
11.600	239.669	26.269	156.731
11.800	243.802	25.478	152.008
12.000	247.934	24.693	147.323
12.100	250.000	24.300	144.981

Table XII (56) Axial Compression Load Capacity of I-Section

ISHB250 @51.0kg/m

Section	*Area mm²*	$r_{min} = r_{yy}$ *mm*	*Buckling Class y-y*
ISHB250 @51.0kg/m	*6495.747*	*54.9*	*c*
Eff Length m	*Sl Ratio*	*D.C.S.* f_{cd} *N/mm²*	*F.A.L. kN*
1.000	18.215	224.536	1458.526
1.200	21.858	221.585	1439.358
1.400	25.501	216.849	1408.595
1.600	29.144	212.113	1377.832
1.800	32.787	207.377	1347.069
2.000	36.430	202.641	1316.306
2.200	40.073	197.891	1285.448
2.400	43.716	192.426	1249.952
2.600	47.359	186.962	1214.456
2.800	51.002	181.497	1178.960
3.000	54.645	176.033	1143.464
3.200	58.288	170.568	1107.969
3.400	61.931	164.911	1071.218
3.600	65.574	159.082	1033.356
3.800	69.217	153.253	995.494
4.000	72.860	147.424	957.632
4.200	76.503	141.596	919.769
4.400	80.146	135.781	882.002

Contd...

4.600	83.789	130.317	846.506
4.800	87.432	124.852	811.010
5.000	91.075	119.495	776.212
5.200	94.718	114.395	743.083
5.400	98.361	109.295	709.953
5.600	102.004	104.515	678.906
5.800	105.647	99.998	649.563
6.000	109.290	95.481	620.220
6.200	112.933	91.403	593.734
6.400	116.576	87.433	567.940
6.600	120.219	83.495	542.359
6.800	123.862	80.070	520.115
7.000	127.505	76.646	497.871
7.200	131.148	73.370	476.596
7.400	134.791	70.420	457.428
7.600	138.434	67.469	438.261
7.800	142.077	64.746	420.577
8.000	145.719	62.196	404.012
8.200	149.362	59.646	387.447
8.400	153.005	57.427	373.030
8.600	156.648	55.277	359.068
8.800	160.291	53.148	345.239
9.000	163.934	51.254	332.934
9.200	167.577	49.360	320.628
9.400	171.220	47.551	308.878
9.600	174.863	45.911	298.229
9.800	178.506	44.272	287.581
10.000	182.149	42.762	277.770
10.200	185.792	41.341	268.541
10.400	189.435	39.920	259.312
10.600	193.078	38.653	251.083
10.800	196.721	37.415	243.037
11.000	200.364	36.191	235.086
11.200	204.007	35.098	227.987
11.400	207.650	34.005	220.887
11.600	211.293	32.951	214.040
11.800	214.936	31.967	207.651
12.000	218.579	30.984	201.262
12.200	222.222	30.089	195.450
12.400	225.865	29.251	190.007
12.600	229.508	28.413	184.564
12.800	233.151	27.638	179.531
13.000	236.794	26.873	174.562

Contd...

13.200	240.437	26.117	169.649
13.400	244.080	25.425	165.153
13.600	247.723	24.733	160.657
13.725	250.000	24.300	157.847

Table XII (57) Axial Compression Load Capacity of I-Section

ISHB250 @54.7kg/m

Section	*Area mm²*	$r_{min} = r_{yy}$ *mm*	*Buckling Class y-y*
ISHB250 @54.7kg/m	*6970.747*	*53.7*	*c*
Eff Length m	*Sl Ratio*	*D.C.S.* f_{cd} *N/mm²*	*F.A.L. kN*
1.000	18.622	224.413	1564.329
1.200	22.346	220.950	1540.185
1.400	26.071	216.108	1506.434
1.600	29.795	211.266	1472.684
1.800	33.520	206.425	1438.934
2.000	37.244	201.583	1405.183
2.200	40.968	196.547	1370.083
2.400	44.693	190.961	1331.140
2.600	48.417	185.374	1292.197
2.800	52.142	179.788	1253.255
3.000	55.866	174.201	1214.312
3.200	59.590	168.615	1175.369
3.400	63.315	162.696	1134.116
3.600	67.039	156.737	1092.577
3.800	70.764	150.778	1051.038
4.000	74.488	144.819	1009.499
4.200	78.212	138.860	967.960
4.400	81.937	133.095	927.771
4.600	85.661	127.508	888.829
4.800	89.385	121.922	849.886
5.000	93.110	116.646	813.111
5.200	96.834	111.432	776.764
5.400	100.559	106.307	741.041
5.600	104.283	101.689	708.848
5.800	108.007	97.071	676.656
6.000	111.732	92.712	646.274
6.200	115.456	88.653	617.976
6.400	119.181	84.593	589.677
6.600	122.905	80.969	564.416
6.800	126.629	77.468	540.012
7.000	130.354	74.013	515.929
7.200	134.078	70.997	494.900
7.400	137.803	67.980	473.871

Contd...

7.600	141.527	65.131	454.012
7.800	145.251	62.524	435.839
8.000	148.976	59.917	417.666
8.200	152.700	57.607	401.563
8.400	156.425	55.409	386.246
8.600	160.149	53.223	371.001
8.800	163.873	51.286	357.501
9.000	167.598	49.349	344.001
9.200	171.322	47.505	331.146
9.400	175.047	45.829	319.463
9.600	178.771	44.153	307.780
9.800	182.495	42.627	297.141
10.000	186.220	41.174	287.016
10.200	189.944	39.722	276.891
10.400	193.669	38.453	268.044
10.600	197.393	37.186	259.217
10.800	201.117	35.965	250.702
11.000	204.842	34.847	242.913
11.200	208.566	33.730	235.124
11.400	212.291	32.682	227.815
11.600	216.015	31.676	220.805
11.800	219.739	30.670	213.796
12.000	223.464	29.803	207.752
12.200	227.188	28.947	201.780
12.400	230.912	28.108	195.936
12.600	234.637	27.326	190.484
12.800	238.361	26.544	185.032
13.000	242.086	25.804	179.871
13.200	245.810	25.096	174.938
13.400	249.534	24.388	170.006
13.425	250.000	24.300	169.389

able XII (58) Axial Compression Load Capacity of I-Section

ISHB300 @58.8kg/m

Section	*Area mm²*	*$r_{min} = r_{yy}$ mm*	*Buckling Class y-y*
ISHB300 @58.8kg/m	*7484.955*	*54.1*	*c*
Eff Length m	*Sl Ratio*	*D.C.S. f_{cd} N/mm²*	*F.A.L. kN*
1.000	18.484	224.455	1680.033
1.200	22.181	221.165	1655.406
1.400	25.878	216.359	1619.434
1.600	29.575	211.553	1583.462
1.800	33.272	206.747	1547.490
2.000	36.969	201.941	1511.518

Contd...

2.200	40.665	197.002	1474.550
2.400	44.362	191.457	1433.044
2.600	48.059	185.911	1391.537
2.800	51.756	180.366	1350.031
3.000	55.453	174.821	1308.525
3.200	59.150	169.275	1267.019
3.400	62.847	163.445	1223.382
3.600	66.543	157.530	1179.109
3.800	70.240	151.616	1134.835
4.000	73.937	145.701	1090.562
4.200	77.634	139.786	1046.289
4.400	81.331	134.004	1003.012
4.600	85.028	128.458	961.505
4.800	88.725	122.913	919.999
5.000	92.421	117.610	880.305
5.200	96.118	112.434	841.566
5.400	99.815	107.259	802.827
5.600	103.512	102.645	768.294
5.800	107.209	98.061	733.982
6.000	110.906	93.613	700.687
6.200	114.603	89.583	670.526
6.400	118.299	85.554	640.365
6.600	121.996	81.823	612.445
6.800	125.693	78.348	586.434
7.000	129.390	74.873	560.424
7.200	133.087	71.800	537.417
7.400	136.784	68.805	515.004
7.600	140.481	65.864	492.986
7.800	144.177	63.276	473.616
8.000	147.874	60.688	454.247
8.200	151.571	58.273	436.171
8.400	155.268	56.092	419.845
8.600	158.965	53.911	403.519
8.800	162.662	51.916	388.588
9.000	166.359	49.994	374.199
9.200	170.055	48.075	359.840
9.400	173.752	46.411	347.388
9.600	177.449	44.748	334.936
9.800	181.146	43.153	322.999
10.000	184.843	41.711	312.207
10.200	188.540	40.270	301.415
10.400	192.237	38.940	291.461
10.600	195.933	37.683	282.053

Contd...

10.800	199.630	36.426	272.645
11.000	203.327	35.302	264.233
11.200	207.024	34.193	255.931
11.400	210.721	33.105	247.792
11.600	214.418	32.107	240.321
11.800	218.115	31.109	232.850
12.000	221.811	30.183	225.921
12.200	225.508	29.333	219.557
12.400	229.205	28.483	213.193
12.600	232.902	27.691	207.263
12.800	236.599	26.914	201.452
13.000	240.296	26.144	195.685
13.200	243.993	25.441	190.428
13.400	247.689	24.739	185.170
13.525	250.000	24.300	181.884

able XII (59) Axial Compression Load Capacity of I-Section

ISHB300 @63.0kg/m

Section	*Area mm^2*	*$r_{min} = r_{yy}$ mm*	*Buckling Class y-y*
ISHB300 @63.0kg/m	*8024.955*	*52.9*	*c*
Eff Length m	*Sl Ratio*	*D.C.S. f_{cd} N/mm^2*	*F.A.L. kN*
1.000	18.904	224.329	1800.229
1.200	22.684	220.510	1769.586
1.400	26.465	215.595	1730.144
1.600	30.246	210.681	1690.702
1.800	34.026	205.766	1651.260
2.000	37.807	200.851	1611.817
2.200	41.588	195.618	1569.827
2.400	45.369	189.947	1524.317
2.600	49.149	184.276	1478.806
2.800	52.930	178.605	1433.296
3.000	56.711	172.934	1387.786
3.200	60.491	167.214	1341.882
3.400	64.272	161.164	1293.337
3.600	68.053	155.115	1244.793
3.800	71.834	149.066	1196.249
4.000	75.614	143.017	1147.705
4.200	79.395	136.968	1099.161
4.400	83.176	131.236	1053.165
4.600	86.957	125.565	1007.655
4.800	90.737	119.968	962.737
5.000	94.518	114.675	920.261
5.200	98.299	109.382	877.784

Contd...

5.400	102.079	104.422	837.978
5.600	105.860	99.733	800.356
5.800	109.641	95.045	762.735
6.000	113.422	90.871	729.232
6.200	117.202	86.750	696.161
6.400	120.983	82.776	664.274
6.600	124.764	79.222	635.754
6.800	128.544	75.668	607.234
7.000	132.325	72.417	581.140
7.200	136.106	69.354	556.565
7.400	139.887	66.292	531.989
7.600	143.667	63.633	510.651
7.800	147.448	60.986	489.413
8.000	151.229	58.475	469.260
8.200	155.009	56.244	451.359
8.400	158.790	54.014	433.458
8.600	162.571	51.963	417.002
8.800	166.352	49.997	401.225
9.000	170.132	48.040	385.522
9.200	173.913	46.339	371.869
9.400	177.694	44.638	358.216
9.600	181.474	43.025	345.273
9.800	185.255	41.550	333.441
10.000	189.036	40.076	321.608
10.200	192.817	38.742	310.906
10.400	196.597	37.457	300.590
10.600	200.378	36.187	290.396
10.800	204.159	35.052	281.294
11.000	207.940	33.918	272.192
11.200	211.720	32.836	263.504
11.400	215.501	31.815	255.312
11.600	219.282	30.794	247.120
11.800	223.062	29.896	239.911
12.000	226.843	29.026	232.933
12.200	230.624	28.169	226.055
12.400	234.405	27.375	219.684
12.600	238.185	26.581	213.312
12.800	241.966	25.826	207.256
13.000	245.747	25.108	201.492
13.200	249.527	24.390	195.727
13.225	250.000	24.300	195.006

able XII (60) Axial Compression Load Capacity of I-Section

ISHB350 @67.4kg/m

Section	*Area mm²*	$r_{min} = r_{yy}$ *mm*	*Buckling Class y-y*
ISHB350 @67.4kg/m	*8591.074*	*53.4*	*b*
Eff Length m	*Sl Ratio*	*D.C.S.* f_{cd} *N/mm²*	*F.A.L. kN*
1.000	18.727	225.255	1935.180
1.200	22.472	222.775	1913.879
1.400	26.217	219.404	1884.920
1.600	29.963	216.034	1855.962
1.800	33.708	212.292	1823.817
2.000	37.453	208.547	1791.641
2.200	41.199	204.562	1757.406
2.400	44.944	200.067	1718.794
2.600	48.689	195.573	1680.182
2.800	52.434	190.835	1639.479
3.000	56.180	185.966	1597.650
3.200	59.925	181.097	1555.821
3.400	63.670	175.494	1507.685
3.600	67.416	169.876	1459.421
3.800	71.161	164.142	1410.159
4.000	74.906	158.150	1358.677
4.200	78.652	152.157	1307.195
4.400	82.397	146.165	1255.713
4.600	86.142	140.172	1204.230
4.800	89.888	134.180	1152.748
5.000	93.633	128.187	1101.266
5.200	97.378	122.195	1049.784
5.400	101.124	116.427	1000.233
5.600	104.869	111.184	955.186
5.800	108.614	105.940	910.139
6.000	112.360	101.098	868.538
6.200	116.105	96.491	828.961
6.400	119.850	91.884	789.385
6.600	123.596	87.853	754.750
6.800	127.341	83.845	720.321
7.000	131.086	80.001	687.292
7.200	134.831	76.555	657.690
7.400	138.577	73.109	628.088
7.600	142.322	69.989	601.279
7.800	146.067	67.067	576.181
8.000	149.813	64.146	551.084
8.200	153.558	61.616	529.349
8.400	157.303	59.107	507.790

Contd...

8.600	161.049	56.692	487.043
8.800	164.794	54.519	468.381
9.000	168.539	52.347	449.719
9.200	172.285	50.358	432.627
9.400	176.030	48.485	416.538
9.600	179.775	46.612	400.450
9.800	183.521	44.986	386.479
10.000	187.266	43.376	372.643
10.200	191.011	41.826	359.329
10.400	194.757	40.440	347.424
10.600	198.502	39.054	335.518
10.800	202.247	37.758	324.385
11.000	205.993	36.522	313.767
11.200	209.738	35.287	303.149
11.400	213.483	34.190	293.728
11.600	217.228	33.104	284.397
11.800	220.974	32.057	275.400
12.000	224.719	31.120	267.356
12.200	228.464	30.184	259.312
12.400	232.210	29.292	251.648
12.600	235.955	28.430	244.247
12.800	239.700	27.569	236.847
13.000	243.446	26.811	230.334
13.200	247.191	26.062	223.899
13.350	250.000	25.500	219.072

Table XII (61) Axial Compression Load Capacity of I-Section

ISHB350 @72.4kg/m

Section	*Area mm²*	*$r_{min} = r_{yy}$ mm*	*Buckling Class y-y*
ISHB350 @72.4kg/m	*9221.074*	*52.2*	*b*
Eff Length m	*Sl Ratio*	*D.C.S. f_{cd} N/mm²*	*F.A.L. kN*
1.000	19.157	225.169	2076.296
1.200	22.989	222.310	2049.940
1.400	26.820	218.862	2018.143
1.600	30.651	215.349	1985.746
1.800	34.483	211.517	1950.416
2.000	38.314	207.686	1915.086
2.200	42.146	203.425	1875.800
2.400	45.977	198.828	1833.404
2.600	49.808	194.230	1791.008
2.800	53.640	189.268	1745.256
3.000	57.471	184.287	1699.327
3.200	61.303	179.046	1650.996

Contd...

3.400	65.134	173.299	1598.002
3.600	68.966	167.552	1545.007
3.800	72.797	161.525	1489.433
4.000	76.628	155.395	1432.905
4.200	80.460	149.264	1376.378
4.400	84.291	143.134	1319.850
4.600	88.123	137.004	1263.322
4.800	91.954	130.874	1206.795
5.000	95.785	124.743	1150.267
5.200	99.617	118.613	1093.740
5.400	103.448	113.172	1043.571
5.600	107.280	107.808	994.110
5.800	111.111	102.633	946.390
6.000	114.943	97.921	902.934
6.200	118.774	93.208	859.478
6.400	122.605	88.912	819.867
6.600	126.437	84.813	782.064
6.800	130.268	80.753	744.632
7.000	134.100	77.228	712.128
7.200	137.931	73.703	679.625
7.400	141.762	70.425	649.397
7.600	145.594	67.437	621.840
7.800	149.425	64.448	594.282
8.000	153.257	61.818	570.028
8.200	157.088	59.251	546.357
8.400	160.920	56.767	523.450
8.600	164.751	54.544	502.958
8.800	168.582	52.322	482.467
9.000	172.414	50.293	463.756
9.200	176.245	48.377	446.092
9.400	180.077	46.467	428.476
9.600	183.908	44.820	413.284
9.800	187.739	43.172	398.092
10.000	191.571	41.619	383.770
10.200	195.402	40.201	370.698
10.400	199.234	38.784	357.626
10.600	203.065	37.489	345.684
10.800	206.897	36.224	334.025
11.000	210.728	34.989	322.635
11.200	214.559	33.878	312.390
11.400	218.391	32.767	302.144
11.600	222.222	31.744	292.718
11.800	226.054	30.787	283.885

Contd...

12.000	229.885	29.829	275.053
12.200	233.716	28.945	266.906
12.400	237.548	28.064	258.780
12.600	241.379	27.224	251.036
12.800	245.211	26.458	243.970
13.000	249.042	25.692	236.904
13.050	250.000	25.500	235.137

Table XII (62) Axial Compression Load Capacity of I-Section

ISHB400 @77.4kg/m

Section	*Area mm²*	$r_{min} = r_{yy}$ *mm*	*Buckling Class y-y*
ISHB400 @77.4kg/m	*9865.890*	*52.6*	*b*
Eff Length m	*Sl Ratio*	*D.C.S.* f_{cd} *N/mm²*	*F.A.L. kN*
1.000	19.011	225.198	2221.776
1.200	22.814	222.468	2194.842
1.400	26.616	219.046	2161.080
1.600	30.418	215.582	2126.906
1.800	34.221	211.779	2089.393
2.000	38.023	207.977	2051.880
2.200	41.825	203.810	2010.766
2.400	45.627	199.247	1965.750
2.600	49.430	194.684	1920.735
2.800	53.232	189.798	1872.531
3.000	57.034	184.856	1823.764
3.200	60.837	179.745	1773.347
3.400	64.639	174.042	1717.077
3.600	68.441	168.338	1660.808
3.800	72.243	162.411	1602.326
4.000	76.046	156.327	1542.305
4.200	79.848	150.243	1482.284
4.400	83.650	144.160	1422.264
4.600	87.452	138.076	1362.243
4.800	91.255	131.992	1302.222
5.000	95.057	125.909	1242.202
5.200	98.859	119.825	1182.181
5.400	102.662	114.274	1127.412
5.600	106.464	108.951	1074.894
5.800	110.266	103.673	1022.823
6.000	114.068	98.996	976.682
6.200	117.871	94.319	930.541
6.400	121.673	89.910	887.041
6.600	125.475	85.841	846.902
6.800	129.278	81.773	806.763

Contd...

7.000	133.080	78.167	771.182
7.200	136.882	74.668	736.671
7.400	140.684	71.266	703.104
7.600	144.487	68.300	673.844
7.800	148.289	65.335	644.584
8.000	152.091	62.599	617.593
8.200	155.894	60.051	592.460
8.400	159.696	57.504	567.326
8.600	163.498	55.271	545.299
8.800	167.300	53.066	523.541
9.000	171.103	50.949	502.654
9.200	174.905	49.048	483.898
9.400	178.707	47.146	465.141
9.600	182.510	45.421	448.118
9.800	186.312	43.786	431.987
10.000	190.114	42.158	415.924
10.200	193.916	40.751	402.044
10.400	197.719	39.344	388.165
10.600	201.521	37.998	374.885
10.800	205.323	36.743	362.506
11.000	209.125	35.489	350.127
11.200	212.928	34.351	338.903
11.400	216.730	33.248	328.024
11.600	220.532	32.167	317.355
11.800	224.335	31.216	307.977
12.000	228.137	30.266	298.599
12.200	231.939	29.354	289.603
12.400	235.741	28.479	280.975
12.600	239.544	27.605	272.347
12.800	243.346	26.831	264.710
13.000	247.148	26.070	257.207
13.150	250.000	25.500	251.580

able XII (63) Axial Compression Load Capacity of I-Section

ISHB400 @82.2kg/m

Section	*Area mm²*	*$r_{min} = r_{yy}$ mm*	*Buckling Class y-y*
ISHB400 @82.2kg/m	*10465.890*	*51.6*	*b*
Eff Length m	*Sl Ratio*	*D.C.S. f_{cd} N/mm²*	*F.A.L. kN*
1.000	19.380	225.124	2356.123
1.200	23.256	222.070	2324.158
1.400	27.132	218.581	2287.649
1.600	31.008	214.992	2250.085
1.800	34.884	211.116	2209.520

Contd...

2.000	38.760	207.240	2168.954
2.200	42.636	202.837	2122.872
2.400	46.512	198.186	2074.193
2.600	50.388	193.496	2025.109
2.800	54.264	188.457	1972.374
3.000	58.140	183.419	1919.639
3.200	62.016	177.977	1862.685
3.400	65.891	172.163	1801.837
3.600	69.767	166.349	1740.989
3.800	73.643	160.171	1676.327
4.000	77.519	153.969	1611.423
4.200	81.395	147.767	1546.518
4.400	85.271	141.566	1481.613
4.600	89.147	135.364	1416.708
4.800	93.023	129.163	1351.804
5.000	96.899	122.961	1286.899
5.200	100.775	116.915	1223.617
5.400	104.651	111.488	1166.825
5.600	108.527	106.062	1110.033
5.800	112.403	101.044	1057.517
6.000	116.279	96.277	1007.622
6.200	120.155	91.534	957.986
6.400	124.031	87.387	914.581
6.600	127.907	83.240	871.176
6.800	131.783	79.360	830.570
7.000	135.659	75.794	793.250
7.200	139.535	72.228	755.929
7.400	143.411	69.140	723.607
7.600	147.287	66.116	691.966
7.800	151.163	63.221	661.663
8.000	155.039	60.624	634.484
8.200	158.915	58.027	607.306
8.400	162.791	55.681	582.755
8.600	166.667	53.433	559.227
8.800	170.543	51.229	536.154
9.000	174.419	49.291	515.871
9.200	178.295	47.353	495.588
9.400	182.171	45.567	476.896
9.600	186.047	43.900	459.453
9.800	189.922	42.233	442.009
10.000	193.798	40.795	426.952
10.200	197.674	39.360	411.942
10.400	201.550	37.988	397.582

Contd...

10.600	205.426	36.709	384.196
10.800	209.302	35.430	370.809
11.000	213.178	34.278	358.753
11.200	217.054	33.154	346.989
11.400	220.930	32.067	335.614
11.600	224.806	31.098	325.473
11.800	228.682	30.129	315.332
12.000	232.558	29.212	305.726
12.200	236.434	28.320	296.396
12.400	240.310	27.438	287.163
12.600	244.186	26.663	279.050
12.800	248.062	25.888	270.937
12.900	250.000	25.500	266.880

able XII (64) Axial Compression Load Capacity of I-Section

ISHB450 @87.2kg/m

Section	Area mm^2	$r_{min} = r_{yy}$ mm	Buckling Class y-y
ISHB450 @87.2kg/m	11114.346	51.8	b
Eff Length m	Sl Ratio	D.C.S. f_{cd} N/mm^2	F.A.L. kN
1.000	19.305	225.139	2502.273
1.200	23.166	222.151	2469.058
1.400	27.027	218.676	2430.437
1.600	30.888	215.112	2390.829
1.800	34.749	211.251	2347.916
2.000	38.610	207.390	2305.004
2.200	42.471	203.035	2256.598
2.400	46.332	198.402	2205.103
2.600	50.193	193.749	2153.394
2.800	54.054	188.730	2097.607
3.000	57.915	183.710	2041.821
3.200	61.776	178.336	1982.087
3.400	65.637	172.544	1917.718
3.600	69.498	166.753	1853.349
3.800	73.359	160.625	1785.247
4.000	77.220	154.448	1716.587
4.200	81.081	148.270	1647.927
4.400	84.942	142.093	1579.267
4.600	88.803	135.915	1510.607
4.800	92.664	129.737	1441.947
5.000	96.525	123.560	1373.287
5.200	100.386	117.459	1305.485
5.400	104.247	112.054	1245.408
5.600	108.108	106.649	1185.330

Contd...

5.800	111.969	101.578	1128.973
6.000	115.830	96.829	1076.191
6.200	119.691	92.080	1023.408
6.400	123.552	87.899	976.942
6.600	127.413	83.768	931.026
6.800	131.274	79.828	887.234
7.000	135.135	76.276	847.754
7.200	138.996	72.724	808.275
7.400	142.857	69.571	773.241
7.600	146.718	66.560	739.769
7.800	150.579	63.612	707.005
8.000	154.440	61.025	678.254
8.200	158.301	58.438	649.503
8.400	162.162	56.046	622.914
8.600	166.023	53.807	598.025
8.800	169.884	51.567	573.135
9.000	173.745	49.627	551.576
9.200	177.606	47.697	530.120
9.400	181.467	45.869	509.805
9.600	185.328	44.209	491.353
9.800	189.189	42.549	472.900
10.000	193.050	41.071	456.482
10.200	196.911	39.643	440.604
10.400	200.772	38.245	425.070
10.600	204.633	36.971	410.909
10.800	208.494	35.697	396.748
11.000	212.355	34.517	383.634
11.200	216.216	33.397	371.189
11.400	220.077	32.281	358.779
11.600	223.938	31.315	348.051
11.800	227.799	30.350	337.323
12.000	231.660	29.418	326.963
12.200	235.521	28.530	317.094
12.400	239.382	27.642	307.224
12.600	243.243	26.851	298.435
12.800	247.104	26.079	289.853
12.950	250.000	25.500	283.416

able XII (65) Axial Compression Load Capacity of I-Section

ISHB450 @92.5kg/m

Section	Area mm^2	$r_{min} = r_{yy}$ mm	Buckling Class y-y
ISHB450 @92.5kg/m	11789.346	50.8	b
Eff Length m	Sl Ratio	D.C.S. f_{cd} N/mm^2	F.A.L. kN
1.000	19.685	225.063	2653.345
1.200	23.622	221.740	2614.171
1.400	27.559	218.197	2572.398
1.600	31.496	214.504	2528.861
1.800	35.433	210.567	2482.446
2.000	39.370	206.630	2436.032
2.200	43.307	202.031	2381.819
2.400	47.244	197.307	2326.121
2.600	51.181	192.465	2269.031
2.800	55.118	187.346	2208.692
3.000	59.055	182.228	2148.353
3.200	62.992	176.512	2080.959
3.400	66.929	170.606	2011.337
3.600	70.866	164.614	1940.693
3.800	74.803	158.315	1866.430
4.000	78.740	152.016	1792.166
4.200	82.677	145.717	1717.903
4.400	86.614	139.417	1643.639
4.600	90.551	133.118	1569.375
4.800	94.488	126.819	1495.112
5.000	98.425	120.520	1420.848
5.200	102.362	114.693	1352.154
5.400	106.299	109.181	1287.174
5.600	110.236	103.709	1222.667
5.800	114.173	98.867	1165.576
6.000	118.110	94.024	1108.486
6.200	122.047	89.509	1055.258
6.400	125.984	85.297	1005.594
6.600	129.921	81.084	955.930
6.800	133.858	77.450	913.089
7.000	137.795	73.828	870.388
7.200	141.732	70.449	830.545
7.400	145.669	67.378	794.342
7.600	149.606	64.307	758.138
7.800	153.543	61.626	726.530
8.000	157.480	58.988	695.432
8.200	161.417	56.478	665.838
8.400	165.354	54.194	638.918

8.600	169.291	51.911	611.997
8.800	173.228	49.886	588.121
9.000	177.165	47.917	564.914
9.200	181.102	46.026	542.616
9.400	185.039	44.333	522.658
9.600	188.976	42.640	502.700
9.800	192.913	41.122	484.802
10.000	196.850	39.665	467.629
10.200	200.787	38.240	450.826
10.400	204.724	36.941	435.510
10.600	208.661	35.642	420.193
10.800	212.598	34.446	406.101
11.000	216.535	33.305	392.641
11.200	220.472	32.182	379.403
11.400	224.409	31.198	367.800
11.600	228.346	30.213	356.196
11.800	232.283	29.275	345.131
12.000	236.220	28.369	334.455
12.200	240.157	27.469	323.836
12.400	244.094	26.681	314.553
12.600	248.031	25.894	305.270
12.700	250.000	25.500	300.628

Factored Axial Compression Load Capacity of Single Channel-Sections

Table XIII (01) Axial Compression Load Capacity of Channel-Section

ISJC100 @ 5.8 kg/m

Section	Area mm^2	$r_{min} = r_{yy}$ mm	Buckling Class y-y
ISJC100 @ 5.8 kg/m	741.318	14.20	c
Eff Length m	Sl Ratio	D.C.S. f_{cd} N/mm^2	F.A.L. kN
1.0	70.423	151.324	112.179
1.1	77.465	140.056	103.826
1.2	84.507	129.239	95.808
1.3	91.549	118.831	88.092
1.4	98.592	108.972	80.783
1.5	105.634	100.014	74.142
1.6	112.676	91.683	67.966
1.7	119.718	84.007	62.276
1.8	126.761	77.345	57.337
1.9	133.803	71.220	52.796
2.0	140.845	65.608	48.637
2.1	147.887	60.679	44.982
2.2	154.930	56.292	41.730
2.3	161.972	52.275	38.752
2.4	169.014	48.613	36.037
2.5	176.056	45.375	33.637
2.6	183.099	42.392	31.426
2.7	190.141	39.652	29.395
2.8	197.183	37.258	27.620
2.9	204.225	35.032	25.970
3.0	211.268	32.958	24.432
3.1	218.310	31.056	23.023
3.2	225.352	29.369	21.772
3.3	232.394	27.797	20.607
3.4	239.437	26.318	19.510
3.5	246.479	24.969	18.510
3.550	250.000	24.3	18.014

Table XIII (02) Axial Compression Load Capacity of Channel-Section

ISJC125 @ 7.9 kg/m

Section	Area mm^2	$r_{min} = r_{yy}$ mm	Buckling Class y-y
ISJC125 @ 7.9 kg/m	1007.410	16.00	c
Eff Length m	Sl Ratio	D.C.S. f_{cd} N/mm^2	F.A.L. kN
1.0	62.500	164.000	165.215
1.1	68.750	154.000	155.141
1.2	75.000	144.000	145.067
1.3	81.250	134.125	135.119
1.4	87.500	124.750	125.674
1.5	93.750	115.750	116.608

Contd...

1.6	100.000	107.000	107.793
1.7	106.250	99.250	99.985
1.8	112.500	91.875	92.556
1.9	118.750	85.063	85.693
2.0	125.000	79.000	79.585
2.1	131.250	73.288	73.831
2.2	137.500	68.225	68.731
2.3	143.750	63.575	64.046
2.4	150.000	59.200	59.639
2.5	156.250	55.513	55.924
2.6	162.500	52.000	52.385
2.7	168.750	48.750	49.111
2.8	175.000	45.850	46.190
2.9	181.250	43.113	43.432
3.0	187.500	40.675	40.976
3.1	193.750	38.425	38.710
3.2	200.000	36.300	36.569
3.3	206.250	34.425	34.680
3.4	212.500	32.625	32.867
3.5	218.750	30.938	31.167
3.6	225.000	29.450	29.668
3.7	231.250	28.038	28.245
3.8	237.500	26.725	26.923
3.9	243.750	25.488	25.676
4.000	250.000	24.300	24.480

able XIII (03) Axial Compression Load Capacity of Channel-Section

ISJC150 @ 9.9 kg/m

Section	*Area mm²*	$r_{min} = r_{yy}$ *mm*	*Buckling Class y-y*
ISJC150 @ 9.9 kg/m	*1265.468*	*17.30*	*c*
Eff Length m	*Sl Ratio*	*D.C.S.* f_{cd} *N/mm²*	*F.A.L. kN*
1.0	57.803	171.295	216.768
1.1	63.584	162.266	205.342
1.2	69.364	153.017	193.639
1.3	75.145	143.769	181.935
1.4	80.925	134.613	170.348
1.5	86.705	125.942	159.376
1.6	92.486	117.520	148.718
1.7	98.266	109.428	138.477
1.8	104.046	101.983	129.056
1.9	109.827	94.815	119.985
2.0	115.607	88.488	111.979
2.1	121.387	82.396	104.269
2.2	127.168	76.962	97.393
2.3	132.948	71.912	91.003

Contd...

2.4	138.728	67.230	85.077
2.5	144.509	63.044	79.780
2.6	150.289	59.029	74.700
2.7	156.069	55.619	70.384
2.8	161.850	52.338	66.232
2.9	167.630	49.332	62.429
3.0	173.410	46.565	58.927
3.1	179.191	43.964	55.635
3.2	184.971	41.661	52.721
3.3	190.751	39.445	49.916
3.4	196.532	37.479	47.429
3.5	202.312	35.606	45.059
3.6	208.092	33.872	42.864
3.7	213.873	32.254	40.817
3.8	219.653	30.694	38.842
3.9	225.434	29.350	37.142
4.0	231.214	28.045	35.490
4.1	236.994	26.831	33.954
4.2	242.775	25.673	32.488
4.3	248.555	24.575	31.098
4.325	250.000	24.300	30.751

Table XIII (04) Axial Compression Load Capacity of Channel-Section

ISJC175 @ 11.2 kg/m

Section	*Area mm²*	$r_{min} = r_{yy}$ *mm*	*Buckling Class y-y*
ISJC175 @ 11.2 kg/m	*1424.468*	*18.80*	*c*
Eff Length m	*Sl Ratio*	*D.C.S.* f_{cd} *N/mm²*	*F.A.L. kN*
1.0	53.191	178.213	253.858
1.1	58.511	170.234	242.493
1.2	63.830	161.872	230.582
1.3	69.149	153.362	218.459
1.4	74.468	144.851	206.336
1.5	79.787	136.340	194.213
1.6	85.106	128.340	182.817
1.7	90.426	120.404	171.512
1.8	95.745	112.957	160.904
1.9	101.064	105.681	150.539
2.0	106.383	99.085	141.144
2.1	111.702	92.745	132.112
2.2	117.021	86.947	123.853
2.3	122.340	81.500	116.094
2.4	127.660	76.500	108.972
2.5	132.979	71.887	102.401
2.6	138.298	67.579	96.264
2.7	143.617	63.668	90.693
2.8	148.936	59.945	85.389

Contd...

2.9	154.255	56.689	80.752
3.0	159.574	53.551	76.282
3.1	164.894	50.755	72.299
3.2	170.213	48.004	68.381
3.3	175.532	45.611	64.971
3.4	180.851	43.268	61.634
3.5	186.170	41.194	58.679
3.6	191.489	39.194	55.830
3.7	196.809	37.385	53.254
3.8	202.128	35.662	50.799
3.9	207.447	34.066	48.526
4.0	212.766	32.553	46.371
4.1	218.085	31.117	44.325
4.2	223.404	29.817	42.473
4.3	228.723	28.594	40.731
4.4	234.043	27.451	39.103
4.5	239.362	26.334	37.512
4.6	244.681	25.311	36.054
4.700	250.000	24.300	34.615

able XIII (05) Axial Compression Load Capacity of Channel-Section

ISJC200 @ 13.9 kg/m

Section	*Area mm²*	$r_{min} = r_{yy}$ *mm*	*Buckling Class y-y*
ISJC200 @ 13.9 kg/m	*1776.671*	*21.80*	*c*
Eff Length m	*Sl Ratio*	*D.C.S.* f_{cd} *N/mm²*	*F.A.L. kN*
1.0	45.872	189.193	336.133
1.1	50.459	182.312	323.908
1.2	55.046	175.431	311.684
1.3	59.633	168.550	299.459
1.4	64.220	161.248	286.484
1.5	68.807	153.908	273.444
1.6	73.394	146.569	260.405
1.7	77.982	139.229	247.365
1.8	82.569	132.147	234.781
1.9	87.156	125.266	222.557
2.0	91.743	118.560	210.642
2.1	96.330	112.138	199.232
2.2	100.917	105.862	188.083
2.3	105.505	100.174	177.977
2.4	110.092	94.500	167.895
2.5	114.679	89.500	159.012
2.6	119.266	84.500	150.129
2.7	123.853	80.078	142.272
2.8	128.440	75.766	134.611
2.9	133.028	71.848	127.650
3.0	137.615	68.132	121.048

Contd...

3.1	142.202	64.659	114.877
3.2	146.789	61.448	109.172
3.3	151.376	58.388	103.736
3.4	155.963	55.682	98.928
3.5	160.550	53.014	94.188
3.6	165.138	50.628	89.950
3.7	169.725	48.243	85.712
3.8	174.312	46.160	82.010
3.9	178.899	44.095	78.343
4.0	183.486	42.240	75.047
4.1	188.073	40.451	71.869
4.2	192.661	38.795	68.927
4.3	197.248	37.236	66.156
4.4	201.835	35.750	63.515
4.5	206.422	34.373	61.070
4.6	211.009	33.028	58.679
4.7	215.596	31.789	56.479
4.8	220.183	30.558	54.291
4.9	224.771	29.503	52.417
5.0	229.358	28.448	50.542
5.1	233.945	27.472	48.808
5.2	238.532	26.508	47.096
5.3	243.119	25.607	45.496
5.4	247.706	24.736	43.947
5.450	250.000	24.300	43.173

Table XIII (06) Axial Compression Load Capacity of Channel-Section

ISLC75 @ 5.7 kg/m

Section	*Area mm²*	$r_{min} = r_{yy}$ *mm*	*Buckling Class y-y*
ISLC75 @ 5.7 kg/m	*726.018*	*12.60*	*c*
Eff Length m	*Sl Ratio*	*D.C.S.* f_{cd} *N/mm²*	*F.A.L. kN*
1.0	79.365	137.016	99.476
1.1	87.302	125.048	90.787
1.2	95.238	113.667	82.524
1.3	103.175	103.063	74.826
1.4	111.111	93.389	67.802
1.5	119.048	84.738	61.521
1.6	126.984	77.135	56.001
1.7	134.921	70.314	51.049
1.8	142.857	64.200	46.610
1.9	150.794	58.732	42.640
2.0	158.730	54.049	39.241
2.1	166.667	49.833	36.180
2.2	174.603	46.029	33.418
2.3	182.540	42.610	30.935
2.4	190.476	39.538	28.705

Contd...

2.5	198.413	36.840	26.746
2.6	206.349	34.395	24.972
2.7	214.286	32.143	23.336
2.8	222.222	30.089	21.845
2.9	230.159	28.267	20.522
3.0	238.095	26.600	19.312
3.1	246.032	25.054	18.190
3.150	250.000	24.300	17.642

able XIII (07) Axial Compression Load Capacity of Channel-Section

ISLC100 @ 7.9 kg/m

Section	*Area mm^2*	*$r_{min} = r_{yy}$ mm*	*Buckling Class y-y*
ISLC100 @ 7.9 kg/m	*1001.718*	*15.70*	*c*
Eff Length m	*Sl Ratio*	*D.C.S. f_{cd} N/mm²*	*F.A.L. kN*
1.0	63.694	162.089	162.368
1.1	70.064	151.898	152.159
1.2	76.433	141.707	141.951
1.3	82.803	131.796	132.023
1.4	89.172	122.242	122.452
1.5	95.541	113.242	113.437
1.6	101.911	104.631	104.810
1.7	108.280	96.732	96.899
1.8	114.650	89.532	89.686
1.9	121.019	82.742	82.884
2.0	127.389	76.755	76.887
2.1	133.758	71.256	71.378
2.2	140.127	66.111	66.224
2.3	146.497	61.652	61.758
2.4	152.866	57.509	57.608
2.5	159.236	53.751	53.843
2.6	165.605	50.385	50.472
2.7	171.975	47.211	47.293
2.8	178.344	44.345	44.421
2.9	184.713	41.762	41.834
3.0	191.083	39.332	39.399
3.1	197.452	37.166	37.230
3.2	203.822	35.154	35.214
3.3	210.191	33.248	33.306
3.4	216.561	31.529	31.583
3.5	222.930	29.926	29.978
3.6	229.299	28.461	28.510
3.7	235.669	27.110	27.156
3.8	242.038	25.813	25.857
3.9	248.408	24.603	24.645
3.925	250.000	24.300	24.342

Table XIII (08) Axial Compression Load Capacity of Channel-Section

ISLC125 @ 10.7 kg/m

Section	Area mm^2	$r_{min} = r_{yy}$ mm	Buckling Class y-y
ISLC125 @ 10.7 kg/m	1367.178	20.50	c
Eff Length m	Sl Ratio	D.C.S. f_{cd} N/mm^2	F.A.L. kN
1.0	48.780	184.829	252.695
1.2	58.537	170.195	232.687
1.4	68.293	154.732	211.546
1.6	78.049	139.122	190.204
1.8	87.805	124.293	169.930
2.0	97.561	110.415	150.956
2.2	107.317	97.927	133.883
2.4	117.073	86.890	118.794
2.6	126.829	77.280	105.656
2.8	136.585	68.966	94.289
3.0	146.341	61.761	84.438
3.2	156.098	55.602	76.018
3.4	165.854	50.256	68.709
3.6	175.610	45.576	62.310
3.8	185.366	41.507	56.748
4.0	195.122	37.959	51.896
4.2	204.878	34.837	47.628
4.4	214.634	32.049	43.816
4.6	224.390	29.590	40.455
4.8	234.146	27.429	37.501
5.0	243.902	25.459	34.806
5.125	250.000	24.300	33.222

Table XIII (09) Axial Compression Load Capacity of Channel-Section

ISLC150 @ 14.4 kg/m

Section	Area mm^2	$r_{min} = r_{yy}$ mm	Buckling Class y-y
ISLC150 @ 14.4 kg/m	1836.011	23.70	c
Eff Length m	Sl Ratio	D.C.S. f_{cd} N/mm^2	F.A.L. kN
1.0	42.194	194.709	357.488
1.2	50.633	182.051	334.247
1.4	59.072	169.392	311.006
1.6	67.511	155.983	286.387
1.8	75.949	142.481	261.597
2.0	84.388	129.418	237.612
2.2	92.827	117.042	214.891
2.4	101.266	105.430	193.571
2.6	109.705	94.966	174.359
2.8	118.143	85.724	157.390
3.0	126.582	77.513	142.314
3.2	135.021	70.233	128.948

Contd...

3.4	143.460	63.778	117.097
3.6	151.899	58.080	106.635
3.8	160.338	53.124	97.537
4.0	168.776	48.736	89.480
4.2	177.215	44.853	82.351
4.4	185.654	41.395	76.002
4.6	194.093	38.308	70.335
4.8	202.532	35.541	65.253
5.0	210.970	33.038	60.658
5.2	219.409	30.759	56.475
5.4	227.848	28.795	52.868
5.6	236.287	26.980	49.535
5.8	244.726	25.302	46.455
5.925	250.000	24.300	44.615

able XIII (10) Axial Compression Load Capacity of Channel-Section

ISLC175 @ 17.6 kg/m

Section	Area mm^2	$r_{min} = r_{yy}$ mm	Buckling Class y-y
ISLC175 @ 17.6 kg/m	2239.977	23.80	c
Eff Length m	Sl Ratio	D.C.S. f_{cd} N/mm^2	F.A.L. kN
1.0	42.017	194.975	436.739
1.2	50.420	182.370	408.504
1.4	58.824	169.765	380.269
1.6	67.227	156.437	350.415
1.8	75.630	142.992	320.298
2.0	84.034	129.950	291.084
2.2	92.437	117.588	263.395
2.4	100.840	105.958	237.343
2.6	109.244	95.538	214.003
2.8	117.647	86.265	193.231
3.0	126.050	78.013	174.746
3.2	134.454	70.692	158.349
3.4	142.857	64.200	143.807
3.6	151.261	58.456	130.941
3.8	159.664	53.498	119.835
4.0	168.067	49.105	109.994
4.2	176.471	45.188	101.221
4.4	184.874	41.699	93.405
4.6	193.277	38.586	86.431
4.8	201.681	35.796	80.182
5.0	210.084	33.277	74.540
5.2	218.487	31.008	69.458
5.4	226.891	29.015	64.993
5.6	235.294	27.188	60.901
5.8	243.697	25.497	57.114
5.950	250.000	24.300	54.431

Table XIII (11) Axial Compression Load Capacity of Channel-Section

ISLC200 @ 20.6 kg/m

Section	Area mm^2	$r_{min} = r_{yy}$ mm	Buckling Class y-y
ISLC200 @ 20.6 kg/m	2622.192	23.70	c
Eff Length m	Sl Ratio	D.C.S. f_{cd} N/mm^2	F.A.L. kN
1.0	42.194	194.709	510.564
1.2	50.633	182.051	477.372
1.4	59.072	169.392	444.179
1.6	67.511	155.983	409.018
1.8	75.949	142.481	373.613
2.0	84.388	129.418	339.358
2.2	92.827	117.042	306.907
2.4	101.266	105.430	276.459
2.6	109.705	94.966	249.020
2.8	118.143	85.724	224.784
3.0	126.582	77.513	203.253
3.2	135.021	70.233	184.164
3.4	143.460	63.778	167.238
3.6	151.899	58.080	152.296
3.8	160.338	53.124	139.303
4.0	168.776	48.736	127.796
4.2	177.215	44.853	117.614
4.4	185.654	41.395	108.545
4.6	194.093	38.308	100.452
4.8	202.532	35.541	93.194
5.0	210.970	33.038	86.632
5.2	219.409	30.759	80.657
5.4	227.848	28.795	75.506
5.6	236.287	26.980	70.746
5.8	244.726	25.302	66.347
5.925	250.000	24.300	63.719

Table XIII (12) Axial Compression Load Capacity of Channel-Section

ISLC225 @ 24.0 kg/m

Section	Area mm^2	$r_{min} = r_{yy}$ mm	Buckling Class y-y
ISLC225 @ 24.0 kg/m	3053.057	26.20	c
Eff Length m	Sl Ratio	D.C.S. f_{cd} N/mm^2	F.A.L. kN
1.0	38.168	200.382	611.777
1.2	45.802	189.298	577.937
1.4	53.435	177.847	542.978
1.6	61.069	166.290	507.693
1.8	68.702	154.076	470.404
2.0	76.336	141.863	433.115
2.2	83.969	130.046	397.037
2.4	91.603	118.756	362.568

Contd...

2.6	99.237	108.069	329.940
2.8	106.870	98.481	300.668
3.0	114.504	89.691	273.831
3.2	122.137	81.691	249.407
3.4	129.771	74.515	227.499
3.6	137.405	68.302	208.531
3.8	145.038	62.673	191.345
4.0	152.672	57.624	175.928
4.2	160.305	53.141	162.243
4.4	167.939	49.172	150.124
4.6	175.573	45.592	139.196
4.8	183.206	42.350	129.296
5.0	190.840	39.415	120.335
5.2	198.473	36.819	112.411
5.4	206.107	34.468	105.233
5.6	213.740	32.290	98.583
5.8	221.374	30.284	92.459
6.0	229.008	28.528	87.098
6.2	236.641	26.905	82.144
6.4	244.275	25.388	77.510
6.550	250.000	24.300	74.189

ble XIII (13) Axial Compression Load Capacity of Channel-Section

ISLC250 @ 28.0 kg/m

Section	*Area mm²*	$r_{min} = r_{yy}$ *mm*	*Buckling Class y-y*
ISLC250 @ 28.0 kg/m	*3564.837*	*28.90*	*c*
Eff Length m	*Sl Ratio*	*D.C.S.* f_{cd} *N/mm²*	*F.A.L. kN*
1.0	34.602	205.017	730.853
1.2	41.522	195.716	697.697
1.4	48.443	185.336	660.691
1.6	55.363	174.955	623.686
1.8	62.284	164.346	585.867
2.0	69.204	153.273	546.395
2.2	76.125	142.201	506.922
2.4	83.045	131.433	468.536
2.6	89.965	121.052	431.530
2.8	96.886	111.360	396.980
3.0	103.806	102.280	364.613
3.2	110.727	93.808	334.410
3.4	117.647	86.265	307.520
3.6	124.567	79.407	283.071
3.8	131.488	73.095	260.571
4.0	138.408	67.489	240.588
4.2	145.329	62.470	222.695
4.4	152.249	57.873	206.308

Contd...

4.6	159.170	53.790	191.752
4.8	166.090	50.133	178.717
5.0	173.010	46.745	166.639
5.2	179.931	43.631	155.538
5.4	186.851	40.928	145.902
5.6	193.772	38.418	136.953
5.8	200.692	36.092	128.663
6.0	207.612	34.016	121.262
6.2	214.533	32.076	114.346
6.4	221.453	30.266	107.892
6.6	228.374	28.674	102.218
6.8	235.294	27.188	96.922
7.0	242.215	25.779	91.899
7.2	249.135	24.464	87.211
7.225	250.000	24.300	86.626

Table XIII (14) Axial Compression Load Capacity of Channel-Section

ISLC300 @ 33.1 kg/m

Section	Area mm^2	$r_{min} = r_{yy}$ mm	Buckling Class y-y
ISLC300 @ 33.1 kg/m	4210.711	28.70	c
Eff Length m	Sl Ratio	D.C.S. f_{cd} N/mm^2	F.A.L. kN
1.0	34.843	204.704	861.949
1.2	41.812	195.282	822.277
1.4	48.780	184.829	778.263
1.6	55.749	174.376	734.248
1.8	62.718	163.652	689.089
2.0	69.686	152.502	642.141
2.2	76.655	141.352	595.192
2.4	83.624	130.564	549.769
2.6	90.592	120.171	506.004
2.8	97.561	110.415	464.924
3.0	104.530	101.383	426.896
3.2	111.498	92.967	391.457
3.4	118.467	85.371	359.473
3.6	125.436	78.591	330.922
3.8	132.404	72.353	304.656
4.0	139.373	66.708	280.888
4.2	146.341	61.761	260.058
4.4	153.310	57.247	241.051
4.6	160.279	53.155	223.821
4.8	167.247	49.531	208.562
5.0	174.216	46.203	194.547
5.2	181.185	43.138	181.642
5.4	188.153	40.420	170.198
5.6	195.122	37.959	159.832

Contd...

5.8	202.091	35.673	150.208
6.0	209.059	33.582	141.405
6.2	216.028	31.672	133.364
6.4	222.997	29.911	125.946
6.6	229.965	28.308	119.197
6.8	236.934	26.844	113.032
7.0	243.902	25.459	107.199
7.175	250.000	24.300	102.320

able XIII (15) Axial Compression Load Capacity of Channel-Section

ISLC350 @ 38.8 kg/m

Section	*Area mm²*	$r_{min} = r_{yy}$ *mm*	*Buckling Class y-y*
ISLC350 @ 38.8 kg/m	*4949.519*	*28.20*	*c*
Eff Length m	*Sl Ratio*	*D.C.S.* f_{cd} *N/mm²*	*F.A.L. kN*
1.0	35.461	203.901	1009.210
1.2	42.553	194.170	961.049
1.4	49.645	183.532	908.395
1.6	56.738	172.894	855.740
1.8	63.830	161.872	801.190
2.0	70.922	150.525	745.026
2.2	78.014	139.177	688.861
2.4	85.106	128.340	635.223
2.6	92.199	117.922	583.657
2.8	99.291	107.993	534.513
3.0	106.383	99.085	490.424
3.2	113.475	90.812	449.476
3.4	120.567	83.167	411.635
3.6	127.660	76.500	378.638
3.8	134.752	70.451	348.699
4.0	141.844	64.909	321.269
4.2	148.936	59.945	296.697
4.4	156.028	55.643	275.407
4.6	163.121	51.677	255.778
4.8	170.213	48.004	237.598
5.0	177.305	44.813	221.802
5.2	184.397	41.885	207.311
5.4	191.489	39.194	193.990
5.6	198.582	36.782	182.055
5.8	205.674	34.598	171.243
6.0	212.766	32.553	161.123
6.2	219.858	30.638	151.645
6.4	226.950	29.001	143.543
6.6	234.043	27.451	135.870
6.8	241.135	25.984	128.610
7.0	248.227	24.637	121.941
7.050	250.000	24.300	120.273

Table XIII (16) Axial Compression Load Capacity of Channel-Section

ISLC400 @ 45.7 kg/m

Section	Area mm^2	$r_{min} = r_{yy}$ mm	Buckling Class y-y
ISLC400 @ 45.7 kg/m	5825.206	28.10	c
Eff Length m	Sl Ratio	D.C.S. f_{cd} N/mm^2	F.A.L. kN
1.0	35.587	203.737	1186.808
1.2	42.705	193.943	1129.758
1.4	49.822	183.267	1067.567
1.6	56.940	172.591	1005.377
1.8	64.057	161.509	940.823
2.0	71.174	150.121	874.486
2.2	78.292	138.733	808.149
2.4	85.409	127.886	744.963
2.6	92.527	117.463	684.244
2.8	99.644	107.498	626.199
3.0	106.762	98.616	574.456
3.2	113.879	90.372	526.435
3.4	120.996	82.763	482.113
3.6	128.114	76.073	443.141
3.8	135.231	70.063	408.129
4.0	142.349	64.556	376.051
4.2	149.466	59.574	347.029
4.4	156.584	55.316	322.225
4.6	163.701	51.375	299.273
4.8	170.819	47.732	278.047
5.0	177.936	44.529	259.390
5.2	185.053	41.629	242.499
5.4	192.171	38.962	226.961
5.6	199.288	36.542	212.865
5.8	206.406	34.378	200.261
6.0	213.523	32.349	188.438
6.2	220.641	30.453	177.393
6.4	227.758	28.816	167.857
6.6	234.875	27.276	158.889
6.8	241.993	25.821	150.415
7.0	249.110	24.469	142.537
7.025	250.000	24.300	141.552

able XIII (17) Axial Compression Load Capacity of Channel-Section

ISMC75 @ 6.8 kg/m

Section	Area mm^2	$r_{min} = r_{yy}$ mm	Buckling Class y-y
ISMC75 @ 6.8 kg/m	867.166	12.10	c
Eff Length m	Sl Ratio	D.C.S. f_{cd} N/mm^2	F.A.L. kN
1.0	82.645	132.033	114.495
1.1	90.909	119.727	103.823
1.2	99.174	108.157	93.790
1.3	107.438	97.777	84.789
1.4	115.702	88.384	76.644
1.5	123.967	79.971	69.348
1.6	132.231	72.493	62.863
1.7	140.496	65.853	57.105
1.8	148.760	60.068	52.089
1.9	157.025	55.055	47.742
2.0	165.289	50.550	43.835
2.1	173.554	46.501	40.324
2.2	181.818	42.891	37.194
2.3	190.083	39.672	34.402
2.4	198.347	36.862	31.965
2.5	206.612	34.317	29.758
2.6	214.876	31.983	27.735
2.7	223.140	29.878	25.909
2.8	231.405	28.005	24.285
2.9	239.669	26.269	22.780
3.0	247.934	24.693	21.413
3.025	250.000	24.300	21.072

ble XIII (18) Axial Compression Load Capacity of Channel-Section

ISMC100 @ 9.2 kg/m

Section	Area mm^2	$r_{min} = r_{yy}$ mm	Buckling Class y-y
ISMC100 @ 9.2 kg/m	1169.835	14.90	c
Eff Length m	Sl Ratio	D.C.S. f_{cd} N/mm^2	F.A.L. kN
1.0	67.114	156.617	183.217
1.1	73.826	145.879	170.655
1.2	80.537	135.195	158.155
1.3	87.248	125.128	146.379
1.4	93.960	115.456	135.065
1.5	100.671	106.168	124.199
1.6	107.383	97.846	114.463
1.7	114.094	90.138	105.446
1.8	120.805	82.943	97.030
1.9	127.517	76.634	89.649
2.0	134.228	70.875	82.912
2.1	140.940	65.542	76.674

Contd...

2.2	147.651	60.844	71.178
2.3	154.362	56.626	66.243
2.4	161.074	52.742	61.699
2.5	167.785	49.252	57.616
2.6	174.497	46.077	53.902
2.7	181.208	43.129	50.454
2.8	187.919	40.511	47.392
2.9	194.631	38.126	44.601
3.0	201.342	35.897	41.994
3.1	208.054	33.884	39.639
3.2	214.765	32.013	37.450
3.3	221.477	30.260	35.400
3.4	228.188	28.717	33.594
3.5	234.899	27.271	31.903
3.6	241.611	25.894	30.292
3.7	248.322	24.619	28.800
3.725	250.000	24.300	28.427

Table XIII (19) Axial Compression Load Capacity of Channel-Section

ISMC125 @ 12.7 kg/m

Section	*Area mm²*	*$r_{min} = r_{yy}$ mm*	*Buckling Class y-y*
ISMC125 @ 12.7 kg/m	*1618.841*	*19.20*	*c*
Eff Length m	*Sl Ratio*	*D.C.S. f_{cd} N/mm²*	*F.A.L. kN*
1.0	52.083	179.875	291.189
1.2	62.500	164.000	265.490
1.4	72.917	147.333	238.509
1.6	83.333	131.000	212.068
1.8	93.750	115.750	187.381
2.0	104.167	101.833	164.852
2.2	114.583	89.604	145.055
2.4	125.000	79.000	127.888
2.6	135.417	69.913	113.177
2.8	145.833	62.117	100.557
3.0	156.250	55.513	89.866
3.2	166.667	49.833	80.672
3.4	177.083	44.913	72.706
3.6	187.500	40.675	65.846
3.8	197.917	37.008	59.911
4.0	208.333	33.800	54.717
4.2	218.750	30.938	50.083
4.4	229.167	28.492	46.123
4.6	239.583	26.288	42.555
4.800	250.000	24.300	39.338

able XIII (20) Axial Compression Load Capacity of Channel-Section

ISMC150 @ 16.4 kg/m

Section	Area mm^2	$r_{min} = r_{yy}$ mm	Buckling Class y-y
ISMC150 @ 16.4 kg/m	2087.905	22.10	c
Eff Length m	Sl Ratio	D.C.S. f_{cd} N/mm^2	F.A.L. kN
1.0	45.249	190.127	396.966
1.2	54.299	176.552	368.624
1.4	63.348	162.643	339.582
1.6	72.398	148.163	309.350
1.8	81.448	133.828	279.420
2.0	90.498	120.303	251.182
2.2	99.548	107.633	224.728
2.4	108.597	96.339	201.147
2.6	117.647	86.265	180.112
2.8	126.697	77.405	161.614
3.0	135.747	69.645	145.413
3.2	144.796	62.843	131.209
3.4	153.846	56.931	118.866
3.6	162.896	51.794	108.141
3.8	171.946	47.224	98.600
4.0	180.995	43.212	90.222
4.2	190.045	39.685	82.858
4.4	199.095	36.608	76.433
4.6	208.145	33.857	70.689
4.8	217.195	31.357	65.471
5.0	226.244	29.164	60.891
5.2	235.294	27.188	56.766
5.4	244.344	25.375	52.980
5.525	250.000	24.300	50.736

ble XIII (21) Axial Compression Load Capacity of Channel-Section

ISMC175 @ 19.1 kg/m

Section	Area mm^2	$r_{min} = r_{yy}$ mm	Buckling Class y-y
ISMC175 @ 19.1 kg/m	2437.999	22.30	c
Eff Length m	Sl Ratio	D.C.S. f_{cd} N/mm^2	F.A.L. kN
1.0	44.843	190.735	465.013
1.2	53.812	177.283	432.215
1.4	62.780	163.552	398.738
1.6	71.749	149.202	363.754
1.8	80.717	134.924	328.944
2.0	89.686	121.471	296.146
2.2	98.655	108.883	265.458
2.4	107.623	97.547	237.820
2.6	116.592	87.415	213.117
2.8	125.561	78.473	191.317

Contd...

3.0	134.529	70.631	172.199
3.2	143.498	63.752	155.426
3.4	152.466	57.745	140.782
3.6	161.435	52.554	128.126
3.8	170.404	47.918	116.825
4.0	179.372	43.883	106.985
4.2	188.341	40.347	98.366
4.4	197.309	37.215	90.730
4.6	206.278	34.417	83.908
4.8	215.247	31.883	77.732
5.0	224.215	29.630	72.239
5.2	233.184	27.631	67.365
5.4	242.152	25.791	62.878
5.575	250.000	24.300	59.243

Table XIII (22) Axial Compression Load Capacity of Channel-Section

ISMC200 @ 22.1 kg/m

Section	*Area mm²*	$r_{min} = r_{yy}$ *mm*	*Buckling Class y-y*
ISMC200 @ 22.1 kg/m	*2821.297*	*22.30*	*c*
Eff Length m	*Sl Ratio*	*D.C.S.* f_{cd} *N/mm²*	*F.A.L. kN*
1.0	44.843	190.735	538.121
1.2	53.812	177.283	500.167
1.4	62.780	163.552	461.428
1.6	71.749	149.202	420.943
1.8	80.717	134.924	380.660
2.0	89.686	121.471	342.705
2.2	98.655	108.883	307.192
2.4	107.623	97.547	275.209
2.6	116.592	87.415	246.623
2.8	125.561	78.473	221.396
3.0	134.529	70.631	199.272
3.2	143.498	63.752	179.862
3.4	152.466	57.745	162.915
3.6	161.435	52.554	148.270
3.8	170.404	47.918	135.192
4.0	179.372	43.883	123.806
4.2	188.341	40.347	113.831
4.4	197.309	37.215	104.994
4.6	206.278	34.417	97.099
4.8	215.247	31.883	89.953
5.0	224.215	29.630	83.596
5.2	233.184	27.631	77.956
5.4	242.152	25.791	72.764
5.6	250.000	24.300	68.558

able XIII (23) Axial Compression Load Capacity of Channel-Section

ISMC225 @ 25.9 kg/m

Section	Area mm^2	$r_{min} = r_{yy}$ mm	Buckling Class y-y
ISMC225 @ 25.9 kg/m	3301.431	23.80	c
Eff Length m	Sl Ratio	D.C.S. f_{cd} N/mm²	F.A.L. kN
1.0	42.017	194.975	643.696
1.2	50.420	182.370	602.081
1.4	58.824	169.765	560.466
1.6	67.227	156.437	516.466
1.8	75.630	142.992	472.077
2.0	84.034	129.950	429.020
2.2	92.437	117.588	388.209
2.4	100.840	105.958	349.813
2.6	109.244	95.538	315.412
2.8	117.647	86.265	284.797
3.0	126.050	78.013	257.553
3.2	134.454	70.692	233.386
3.4	142.857	64.200	211.952
3.6	151.261	58.456	192.989
3.8	159.664	53.498	176.621
4.0	168.067	49.105	162.117
4.2	176.471	45.188	149.186
4.4	184.874	41.699	137.667
4.6	193.277	38.586	127.388
4.8	201.681	35.796	118.177
5.0	210.084	33.277	109.863
5.2	218.487	31.008	102.372
5.4	226.891	29.015	95.791
5.6	235.294	27.188	89.760
5.8	243.697	25.497	84.178
5.950	250.000	24.300	80.225

ble XIII (24) Axial Compression Load Capacity of Channel-Section

ISMC250 @ 30.4 kg/m

Section	Area mm^2	$r_{min} = r_{yy}$ mm	Buckling Class y-y
ISMC250 @ 30.4 kg/m	3866.931	23.80	c
Eff Length m	Sl Ratio	D.C.S. f_{cd} N/mm²	F.A.L. kN
1.0	42.017	194.975	753.954
1.2	50.420	182.370	705.211
1.4	58.824	169.765	656.468
1.6	67.227	156.437	604.931
1.8	75.630	142.992	552.939
2.0	84.034	129.950	502.506
2.2	92.437	117.588	454.706
2.4	100.840	105.958	409.732

Contd...

2.6	109.244	95.538	369.438
2.8	117.647	86.265	333.580
3.0	126.050	78.013	301.669
3.2	134.454	70.692	273.363
3.4	142.857	64.200	248.257
3.6	151.261	58.456	226.046
3.8	159.664	53.498	206.874
4.0	168.067	49.105	189.886
4.2	176.471	45.188	174.740
4.4	184.874	41.699	161.248
4.6	193.277	38.586	149.208
4.8	201.681	35.796	138.420
5.0	210.084	33.277	128.681
5.2	218.487	31.008	119.907
5.4	226.891	29.015	112.199
5.6	235.294	27.188	105.135
5.8	243.697	25.497	98.597
5.950	250.000	24.300	93.966

Table XIII (25) Axial Compression Load Capacity of Channel-Section

ISMC300 @ 35.8 kg/m

Section	*Area mm²*	$r_{min} = r_{yy}$ *mm*	*Buckling Class y-y*
ISMC300 @ 35.8 kg/m	*4563.707*	*26.10*	*c*
Eff Length m	*Sl Ratio*	*D.C.S.* f_{cd} *N/mm²*	*F.A.L. kN*
1.0	38.314	200.192	913.616
1.2	45.977	189.034	862.698
1.4	53.640	177.540	810.242
1.6	61.303	165.916	757.191
1.8	68.966	153.655	701.237
2.0	76.628	141.395	645.284
2.2	84.291	129.563	591.289
2.4	91.954	118.264	539.724
2.6	99.617	107.536	490.765
2.8	107.280	97.973	447.121
3.0	114.943	89.213	407.140
3.2	122.605	81.251	370.806
3.4	130.268	74.083	338.092
3.6	137.931	67.876	309.766
3.8	145.594	62.284	284.247
4.0	153.257	57.279	261.403
4.2	160.920	52.822	241.063
4.4	168.582	48.837	222.879
4.6	176.245	45.290	206.689
4.8	183.908	42.076	192.022
5.0	191.571	39.166	178.742

Contd...

5.2	199.234	36.561	166.852
5.4	206.897	34.231	156.220
5.6	214.559	32.069	146.353
5.8	222.222	30.089	137.317
6.0	229.885	28.326	129.274
6.2	237.548	26.715	121.919
6.4	245.211	25.210	115.051
6.525	250.000	24.300	110.898

able XIII (26) Axial Compression Load Capacity of Channel-Section

ISMC350 @ 42.1 kg/m

Section	*Area mm²*	$r_{min} = r_{yy}$ *mm*	*Buckling Class y-y*
ISMC350 @ 42.1 kg/m	*5365.506*	*28.30*	*c*
Eff Length m	*Sl Ratio*	*D.C.S.* f_{cd} *N/mm²*	*F.A.L. kN*
1.0	35.336	204.064	1094.904
1.2	42.403	194.396	1043.032
1.4	49.470	183.795	986.153
1.6	56.537	173.194	929.275
1.8	63.604	162.233	870.463
2.0	70.671	150.926	809.793
2.2	77.739	139.618	749.123
2.4	84.806	128.792	691.032
2.6	91.873	118.378	635.158
2.8	98.940	108.484	582.072
3.0	106.007	99.551	534.143
3.2	113.074	91.249	489.598
3.4	120.141	83.567	448.380
3.6	127.208	76.924	412.736
3.8	134.276	70.837	380.075
4.0	141.343	65.260	350.153
4.2	148.410	60.313	323.610
4.4	155.477	55.969	300.300
4.6	162.544	51.977	278.883
4.8	169.611	48.302	259.165
5.0	176.678	45.095	241.956
5.2	183.746	42.139	226.098
5.4	190.813	39.424	211.528
5.6	197.880	37.021	198.636
5.8	204.947	34.816	186.805
6.0	212.014	32.756	175.753
6.2	219.081	30.848	165.515
6.4	226.148	29.186	156.597
6.6	233.216	27.625	148.221
6.8	240.283	26.146	140.288
7.0	247.350	24.804	133.083
7.075	250.000	24.300	130.382

Table XIII (27) Axial Compression Load Capacity of Channel-Section

ISMC400 @ 49.4 kg/m

Section	Area mm²	$r_{min} = r_{yy}$ mm	Buckling Class y-y
ISMC400 @ 49.4 kg/m	6293.326	28.30	c
Eff Length m	Sl Ratio	D.C.S. f_{cd} N/mm²	F.A.L. kN
1.0	35.336	204.064	1284.239
1.2	42.403	194.396	1223.396
1.4	49.470	183.795	1156.682
1.6	56.537	173.194	1089.968
1.8	63.604	162.233	1020.987
2.0	70.671	150.926	949.825
2.2	77.739	139.618	878.664
2.4	84.806	128.792	810.527
2.6	91.873	118.378	744.992
2.8	98.940	108.484	682.726
3.0	106.007	99.551	626.508
3.2	113.074	91.249	574.260
3.4	120.141	83.567	525.915
3.6	127.208	76.924	484.108
3.8	134.276	70.837	445.799
4.0	141.343	65.260	410.703
4.2	148.410	60.313	379.570
4.4	155.477	55.969	352.228
4.6	162.544	51.977	327.108
4.8	169.611	48.302	303.981
5.0	176.678	45.095	283.796
5.2	183.746	42.139	265.196
5.4	190.813	39.424	248.106
5.6	197.880	37.021	232.984
5.8	204.947	34.816	219.108
6.0	212.014	32.756	206.145
6.2	219.081	30.848	194.137
6.4	226.148	29.186	183.676
6.6	233.216	27.625	173.851
6.8	240.283	26.146	164.547
7.0	247.350	24.804	156.097
7.075	250.000	24.300	152.928

Factored Axial Compression Load Capacity of Single Tee-Sections

Table XIV (01) Axial Compression Load Capacity of T-Section

ISNT 20 @ 0.9 kg/m

Section	*Area mm²*	*$r_{min} = r_{yy}$ mm*	*Buckling Class y-y*
ISNT 20 @ 0.9 kg/m	*113.884*	*3.90*	*c*
Eff Length m	*Sl Ratio*	*D.C.S. f_{cd} N/mm²*	*F.A.L. kN*
0.10	25.641	216.667	24.675
0.15	38.462	200.000	22.777
0.20	51.282	181.077	20.622
0.25	64.103	161.436	18.385
0.30	76.923	140.923	16.049
0.35	89.744	121.385	13.824
0.40	102.564	103.821	11.824
0.45	115.385	88.731	10.105
0.50	128.205	75.987	8.654
0.55	141.026	65.482	7.457
0.60	153.846	56.931	6.484
0.65	166.667	49.833	5.675
0.70	179.487	43.831	4.992
0.75	192.308	38.915	4.432
0.80	205.128	34.762	3.959
0.85	217.949	31.154	3.548
0.90	230.769	28.138	3.205
0.95	243.590	25.518	2.906
0.975	250.000	24.300	2.767

Table XIV (02) Axial Compression Load Capacity of T-Section

ISNT 30 @ 1.4 kg/m

Section	*Area mm²*	*$r_{min} = r_{yy}$ mm*	*Buckling Class y-y*
ISNT 30 @ 1.4 kg/m	*176.253*	*5.70*	*c*
Eff Length m	*Sl Ratio*	*D.C.S. f_{cd} N/mm²*	*F.A.L. kN*
0.10	17.544	224.737	39.611
0.20	35.088	204.386	36.024
0.30	52.632	179.053	31.559
0.40	70.175	151.719	26.741
0.50	87.719	124.421	21.930
0.60	105.263	100.474	17.709
0.70	122.807	81.061	14.287
0.80	140.351	65.954	11.625
0.90	157.895	54.542	9.613
1.00	175.439	45.653	8.046
1.10	192.982	38.686	6.819
1.20	210.526	33.158	5.844
1.30	228.070	28.744	5.066
1.40	245.614	25.133	4.430
1.425	250.000	24.300	4.283

Table XIV (03) Axial Compression Load Capacity of T-Section

ISNT 40 @ 3.5 kg/m

Section	Area mm²	$r_{min} = r_{yy}$ mm	Buckling Class y-y
ISNT 40 @ 3.5 kg/m	449.871	8.20	c
Eff Length m	Sl Ratio	D.C.S. f_{cd} N/mm²	F.A.L. kN
0.10	12.195	226.341	101.825
0.20	24.390	218.293	98.204
0.30	36.585	202.439	91.072
0.40	48.780	184.829	83.149
0.50	60.976	166.439	74.876
0.60	73.171	146.927	66.098
0.70	85.366	127.951	57.562
0.80	97.561	110.415	49.672
0.90	109.756	94.902	42.694
1.00	121.951	81.866	36.829
1.10	134.146	70.941	31.915
1.20	146.341	61.761	27.785
1.30	158.537	54.163	24.367
1.40	170.732	47.771	21.491
1.50	182.927	42.459	19.101
1.60	195.122	37.959	17.076
1.70	207.317	34.105	15.343
1.80	219.512	30.732	13.825
1.90	231.707	27.941	12.570
2.00	243.902	25.459	11.453
2.050	250.000	24.300	10.932

Table XIV (04) Axial Compression Load Capacity of T-Section

ISNT 50 @ 4.5 kg/m

Section	Area mm²	$r_{min} = r_{yy}$ mm	Buckling Class y-y
ISNT 50 @ 4.5 kg/m	572.241	10.20	c
Eff Length m	Sl Ratio	D.C.S. f_{cd} N/mm²	F.A.L. kN
0.20	19.608	224.118	128.249
0.30	29.412	211.765	121.180
0.40	39.216	199.020	113.887
0.50	49.020	184.471	105.562
0.60	58.824	169.765	97.146
0.70	68.627	154.196	88.237
0.80	78.431	138.510	79.261
0.90	88.235	123.647	70.756
1.00	98.039	109.745	62.801
1.10	107.843	97.275	55.664
1.20	117.647	86.265	49.364
1.30	127.451	76.696	43.889
1.40	137.255	68.424	39.155

Contd...

1.50	147.059	61.259	35.055
1.60	156.863	55.151	31.560
1.70	166.667	49.833	28.517
1.80	176.471	45.188	25.859
1.90	186.275	41.153	23.549
2.00	196.078	37.633	21.535
2.10	205.882	34.535	19.763
2.20	215.686	31.765	18.177
2.30	225.490	29.337	16.788
2.40	235.294	27.188	15.558
2.50	245.098	25.231	14.438
2.550	250.000	24.300	13.905

Table XIV (05) Axial Compression Load Capacity of T-Section

ISNT 60 @ 5.4 kg/m

Section	*Area mm²*	$r_{min} = r_{yy}$ *mm*	*Buckling Class y-y*
ISNT 60 @ 5.4 kg/m	*693.065*	*12.10*	*c*
Eff Length m	*Sl Ratio*	*D.C.S.* f_{cd} *N/mm²*	*F.A.L. kN*
0.20	16.529	225.041	155.968
0.30	24.793	217.769	150.928
0.40	33.058	207.025	143.482
0.50	41.322	196.017	135.852
0.60	49.587	183.620	127.261
0.70	57.851	171.223	118.669
0.80	66.116	158.215	109.653
0.90	74.380	144.992	100.489
1.00	82.645	132.033	91.508
1.10	90.909	119.727	82.979
1.20	99.174	108.157	74.960
1.30	107.438	97.777	67.766
1.40	115.702	88.384	61.256
1.50	123.967	79.971	55.425
1.60	132.231	72.493	50.242
1.70	140.496	65.853	45.640
1.80	148.760	60.068	41.631
1.90	157.025	55.055	38.157
2.00	165.289	50.550	35.034
2.10	173.554	46.501	32.228
2.20	181.818	42.891	29.726
2.30	190.083	39.672	27.495
2.40	198.347	36.862	25.548
2.50	206.612	34.317	23.784
2.60	214.876	31.983	22.167
2.70	223.140	29.878	20.707
2.80	231.405	28.005	19.409

Contd...

2.90	239.669	26.269	18.206
3.00	247.934	24.693	17.114
3.025	250.000	24.300	16.841

able XIV (06) Axial Compression Load Capacity of Channel-Section

ISNT 80 @ 9.6 kg/m

Section	*Area mm^2*	*$r_{min} = r_{yy}$ mm*	*Buckling Class y-y*
ISNT 80 @ 9.6 kg/m	*1229.906*	*16.20*	*c*
Eff Length m	*Sl Ratio*	*D.C.S. f_{cd} N/mm²*	*F.A.L. kN*
0.20	12.346	226.296	278.323
0.40	24.691	217.901	267.998
0.60	37.037	201.852	248.259
0.80	49.383	183.926	226.212
1.00	61.728	165.235	203.223
1.20	74.074	145.481	178.929
1.40	86.420	126.370	155.424
1.60	98.765	108.728	133.726
1.80	111.111	93.389	114.860
2.00	123.457	80.451	98.947
2.20	135.802	69.600	85.602
2.40	148.148	60.496	74.405
2.60	160.494	53.043	65.238
2.80	172.840	46.822	57.587
3.00	185.185	41.578	51.137
3.20	197.531	37.140	45.678
3.40	209.877	33.337	41.001
3.60	222.222	30.089	37.007
3.80	234.568	27.341	33.627
4.00	246.914	24.886	30.608
4.050	250.000	24.300	29.887

ble XIV (07) Axial Compression Load Capacity of Channel-Section

ISNT 100 @ 15.0 kg/m

Section	*Area mm^2*	*$r_{min} = r_{yy}$ mm*	*Buckling Class y-y*
ISNT 100 @ 15.0 kg/m	*1918.542*	*20.50*	*c*
Eff Length m	*Sl Ratio*	*D.C.S. f_{cd} N/mm²*	*F.A.L. kN*
1.00	48.780	184.829	354.603
1.20	58.537	170.195	326.527
1.40	68.293	154.732	296.859
1.60	78.049	139.122	266.911
1.80	87.805	124.293	238.461
2.00	97.561	110.415	211.835
2.20	107.317	97.927	187.877
2.40	117.073	86.890	166.703
2.60	126.829	77.280	148.266

Contd...

2.80	136.585	68.966	132.314
3.00	146.341	61.761	118.491
3.20	156.098	55.602	106.676
3.40	165.854	50.256	96.418
3.60	175.610	45.576	87.439
3.80	185.366	41.507	79.634
4.00	195.122	37.959	72.825
4.20	204.878	34.837	66.835
4.40	214.634	32.049	61.487
4.60	224.390	29.590	56.770
4.80	234.146	27.429	52.624
5.00	243.902	25.459	48.843
5.125	250.000	24.300	46.621

Table XIV (08) Axial Compression Load Capacity of Channel-Section

ISNT 150 @ 22.8 kg/m

Section	*Area mm²*	*$r_{min} = r_{yy}$ mm*	*Buckling Class y-y*
ISNT 150 @ 22.8 kg/m	*2921.014*	*30.30*	*c*
Eff Length m	*Sl Ratio*	*D.C.S. f_{cd} N/mm²*	*F.A.L. kN*
1.00	33.003	207.096	604.930
1.20	39.604	198.515	579.865
1.40	46.205	188.693	551.175
1.60	52.805	178.792	522.254
1.80	59.406	168.891	493.333
2.00	66.007	158.389	462.658
2.20	72.607	147.828	431.809
2.40	79.208	137.267	400.960
2.60	85.809	127.287	371.808
2.80	92.409	117.627	343.590
3.00	99.010	108.386	316.597
3.20	105.611	100.043	292.227
3.40	112.211	92.190	269.288
3.60	118.812	84.995	248.272
3.80	125.413	78.612	229.627
4.00	132.013	72.669	212.268
4.20	138.614	67.323	196.651
4.40	145.215	62.550	182.709
4.60	151.815	58.129	169.796
4.80	158.416	54.235	158.420
5.00	165.017	50.691	148.070
5.20	171.617	47.372	138.375
5.40	178.218	44.402	129.699
5.60	184.818	41.721	121.867
5.80	191.419	39.217	114.555
6.00	198.020	36.973	107.999

Contd...

6.20	204.620	34.914	101.984
6.40	211.221	32.970	96.307
6.60	217.822	31.188	91.101
6.80	224.422	29.583	86.412
7.00	231.023	28.085	82.037
7.20	237.624	26.699	77.988
7.40	244.224	25.397	74.186
7.575	250.000	24.300	70.981

able XIV (09) Axial Compression Load Capacity of Channel-Section

ISHT 75 @ 15.3 kg/m

Section	*Area mm²*	$r_{min} = r_{zz}$ *mm*	*Buckling Class y-y*
ISHT 75 @ 15.3 kg/m	*1948.874*	*22.20*	*c*
Eff Length m	*Sl Ratio*	*D.C.S.* f_{cd} *N/mm²*	*F.A.L. kN*
1.00	45.045	190.432	371.129
1.20	54.054	176.919	344.793
1.40	63.063	163.099	317.860
1.60	72.072	148.685	289.768
1.80	81.081	134.378	261.887
2.00	90.090	120.874	235.568
2.20	99.099	108.261	210.988
2.40	108.108	96.946	188.935
2.60	117.117	86.842	169.245
2.80	126.126	77.941	151.898
3.00	135.135	70.141	136.695
3.20	144.144	63.299	123.362
3.40	153.153	57.340	111.748
3.60	162.162	52.176	101.684
3.80	171.171	47.573	92.714
4.00	180.180	43.530	84.834
4.20	189.189	40.016	77.987
4.40	198.198	36.913	71.938
4.60	207.207	34.138	66.530
4.80	216.216	31.622	61.627
5.00	225.225	29.398	57.293
5.20	234.234	27.411	53.420
5.40	243.243	25.584	49.860
5.550	250.000	24.300	47.358

Table XIV (10) Axial Compression Load Capacity of Channel-Section

ISHT 100 @ 20.0 kg/m

Section	Area mm²	$r_{min} = r_{zz}$ mm	Buckling Class y-y
ISHT 100 @ 20.0 kg/m	2547.215	27.60	c
Eff Length m	Sl Ratio	D.C.S. f_{cd} N/mm²	F.A.L. kN
1.00	36.232	202.899	516.826
1.20	43.478	192.783	491.059
1.40	50.725	181.913	463.372
1.60	57.971	171.043	435.685
1.80	65.217	159.652	406.669
2.00	72.464	148.058	377.136
2.20	79.710	136.464	347.603
2.40	86.957	125.565	319.842
2.60	94.203	115.116	293.225
2.80	101.449	105.203	267.974
3.00	108.696	96.217	245.086
3.20	115.942	88.123	224.469
3.40	123.188	80.703	205.568
3.60	130.435	73.948	188.361
3.80	137.681	68.078	173.410
4.00	144.928	62.751	159.840
4.20	152.174	57.917	147.528
4.40	159.420	53.642	136.638
4.60	166.667	49.833	126.936
4.80	173.913	46.339	118.036
5.00	181.159	43.148	109.907
5.20	188.406	40.322	102.708
5.40	195.652	37.778	96.229
5.60	202.899	35.430	90.249
5.80	210.145	33.261	84.723
6.00	217.391	31.304	79.739
6.20	224.638	29.533	75.228
6.40	231.884	27.904	71.078
6.60	239.130	26.383	67.202
6.80	246.377	24.988	63.651
6.900	250.000	24.300	61.897

Table XIV (11) Axial Compression Load Capacity of Channel-Section

ISHT 125 @ 27.4 kg/m

Section	Area mm²	$r_{min} = r_{zz}$ mm	Buckling Class y-y
ISHT 125 @ 27.4 kg/m	3485.373	34.50	c
Eff Length m	Sl Ratio	D.C.S. f_{cd} N/mm²	F.A.L. kN
1.00	28.986	212.319	740.010
1.20	34.783	204.783	713.744
1.40	40.580	197.130	687.073

Contd...

1.60	46.377	188.435	656.766
1.80	52.174	179.739	626.458
2.00	57.971	171.043	596.150
2.20	63.768	161.971	564.529
2.40	69.565	152.696	532.201
2.60	75.362	143.420	499.873
2.80	81.159	134.261	467.949
3.00	86.957	125.565	437.642
3.20	92.754	117.145	408.294
3.40	98.551	109.029	380.007
3.60	104.348	101.609	354.144
3.80	110.145	94.442	329.166
4.00	115.942	88.123	307.142
4.20	121.739	82.065	286.028
4.40	127.536	76.616	267.035
4.60	133.333	71.600	249.553
4.80	139.130	66.904	233.187
5.00	144.928	62.751	218.710
5.20	150.725	58.772	204.844
5.40	156.522	55.352	192.923
5.60	162.319	52.094	181.568
5.80	168.116	49.080	171.061
6.00	173.913	46.339	161.509
6.20	179.710	43.730	152.417
6.40	185.507	41.452	144.476
6.60	191.304	39.257	136.824
6.80	197.101	37.286	129.954
7.00	202.899	35.430	123.488
7.20	208.696	33.691	117.427
7.40	214.493	32.087	111.835
7.60	220.290	30.533	106.420
7.80	226.087	29.200	101.773
8.00	231.884	27.904	97.257
8.20	237.681	26.687	93.014
8.40	243.478	25.539	89.013
8.60	249.275	24.438	85.174
8.625	250.000	24.300	84.695

ble XIV (12) Axial Compression Load Capacity of Channel-Section

ISHT 150 @ 29.4 kg/m

Section	*Area mm²*	*$r_{min} = r_{zz}$ mm*	*Buckling Class y-y*
ISHT 150 @ 29.4 kg/m	*3742.477*	*39.20*	*c*
Eff Length m	*Sl Ratio*	*D.C.S. f_{cd} N/mm²*	*F.A.L. kN*
1.00	25.510	216.837	811.507
1.20	30.612	210.204	786.684
1.40	35.714	203.571	761.861

Contd...

1.60	40.816	196.776	736.428
1.80	45.918	189.122	707.786
2.00	51.020	181.469	679.145
2.20	56.122	173.816	650.504
2.40	61.224	166.041	621.404
2.60	66.327	157.878	590.853
2.80	71.429	149.714	560.302
3.00	76.531	141.551	529.751
3.20	81.633	133.551	499.812
3.40	86.735	125.898	471.170
3.60	91.837	118.429	443.216
3.80	96.939	111.286	416.484
4.00	102.041	104.469	390.974
4.20	107.143	98.143	367.297
4.40	112.245	92.153	344.881
4.60	117.347	86.592	324.068
4.80	122.449	81.398	304.630
5.00	127.551	76.602	286.681
5.20	132.653	72.151	270.024
5.40	137.755	68.018	254.557
5.60	142.857	64.200	240.267
5.80	147.959	60.629	226.901
6.00	153.061	57.394	214.795
6.20	158.163	54.384	203.530
6.40	163.265	51.602	193.119
6.60	168.367	48.949	183.190
6.80	173.469	46.539	174.170
7.00	178.571	44.243	165.578
7.20	183.673	42.167	157.810
7.40	188.776	40.178	150.364
7.60	193.878	38.382	143.642
7.80	198.980	36.647	137.150
8.00	204.082	35.076	131.269
8.20	209.184	33.545	125.541
8.40	214.286	32.143	120.294
8.60	219.388	30.765	115.138
8.80	224.490	29.567	110.655
9.00	229.592	28.394	106.263
9.20	234.694	27.314	102.223
9.40	239.796	26.243	98.213
9.60	244.898	25.269	94.570
9.800	250.000	24.300	90.942

Table XIV (13) Axial Compression Load Capacity of Channel-Section

ISST 100 @ 8.1 kg/m

Section	Area mm²	$r_{min} = r_{yy}$ mm	Buckling Class y-y
ISST 100 @ 8.1 kg/m	1036.755	9.60	c
Eff Length m	Sl Ratio	D.C.S. f_{cd} N/mm²	F.A.L. kN
1.00	104.167	101.833	105.576
1.10	114.583	89.604	92.898
1.20	125.000	79.000	81.904
1.30	135.417	69.913	72.482
1.40	145.833	62.117	64.400
1.50	156.250	55.513	57.553
1.60	166.667	49.833	51.665
1.70	177.083	44.913	46.563
1.80	187.500	40.675	42.170
1.90	197.917	37.008	38.369
2.00	208.333	33.800	35.042
2.10	218.750	30.938	32.075
2.20	229.167	28.492	29.539
2.30	239.583	26.288	27.254
2.400	250.000	24.300	25.193

Table XIV (14) Axial Compression Load Capacity of Channel-Section

ISST 150 @ 15.7 kg/m

Section	Area mm²	$r_{min} = r_{yy}$ mm	Buckling Class y-y
ISST 150 @ 15.7 kg/m	1995.875	13.60	c
Eff Length m	Sl Ratio	D.C.S. f_{cd} N/mm²	F.A.L. kN
1.00	73.529	146.353	292.102
1.10	80.882	134.676	268.797
1.20	88.235	123.647	246.784
1.30	95.588	113.176	225.886
1.40	102.941	103.353	206.280
1.50	110.294	94.279	188.170
1.60	117.647	86.265	172.174
1.70	125.000	79.000	157.674
1.80	132.353	72.394	144.490
1.90	139.706	66.438	132.602
2.00	147.059	61.259	122.265
2.10	154.412	56.597	112.961
2.20	161.765	52.382	104.549
2.30	169.118	48.559	96.917
2.40	176.471	45.188	90.190
2.50	183.824	42.109	84.044
2.60	191.176	39.300	78.438
2.70	198.529	36.800	73.448

Contd...

2.80	205.882	34.535	68.928
2.90	213.235	32.426	64.719
3.00	220.588	30.465	60.804
3.10	227.941	28.774	57.428
3.20	235.294	27.188	54.264
3.30	242.647	25.697	51.288
3.400	250.000	24.300	48.500

Table XIV (15) Axial Compression Load Capacity of Channel-Section

ISST 200 @ 28.4 kg/m

Section	*Area mm²*	$r_{min} = r_{yy}$ *mm*	*Buckling Class y-y*
ISST 200 @ 28.4 kg/m	*3621.521*	*31.50*	*c*
Eff Length m	*Sl Ratio*	*D.C.S.* f_{cd} *N/mm²*	*F.A.L. kN*
1.00	31.746	208.730	755.921
1.20	38.095	200.476	726.029
1.40	44.444	191.333	692.918
1.60	50.794	181.810	658.427
1.80	57.143	172.286	623.936
2.00	63.492	162.413	588.181
2.20	69.841	152.254	551.391
2.40	76.190	142.095	514.601
2.60	82.540	132.190	478.731
2.80	88.889	122.667	444.240
3.00	95.238	113.667	411.646
3.20	101.587	105.032	380.375
3.40	107.937	97.159	351.862
3.60	114.286	89.929	325.678
3.80	120.635	83.103	300.960
4.00	126.984	77.135	279.346
4.20	133.333	71.600	259.301
4.40	139.683	66.457	240.676
4.60	146.032	61.978	224.454
4.80	152.381	57.795	209.307
5.00	158.730	54.049	195.740
5.20	165.079	50.659	183.462
5.40	171.429	47.457	171.867
5.60	177.778	44.600	161.520
5.80	184.127	41.990	152.069
6.00	190.476	39.538	143.188
6.20	196.825	37.379	135.370
6.40	203.175	35.348	128.012
6.60	209.524	33.443	121.114
6.80	215.873	31.714	114.854
7.00	222.222	30.089	108.968
7.20	228.571	28.629	103.679

Contd...

7.40	234.921	27.267	98.747
7.60	241.270	25.959	94.010
7.80	247.619	24.752	89.641
7.875	250.000	24.300	88.003

able XIV (16) Axial Compression Load Capacity of Channel-Section

ISST 250 @ 37.5 kg/m

Section	*Area mm²*	*$r_{min} = r_{yy}$ mm*	*Buckling Class y-y*
ISST 250 @ 37.5 kg/m	*4774.910*	*33.40*	*c*
Eff Length m	*Sl Ratio*	*D.C.S. f_{cd} N/mm²*	*F.A.L. kN*
1.00	29.940	211.078	1007.878
1.20	35.928	203.293	970.708
1.40	41.916	195.126	931.708
1.60	47.904	186.144	888.819
1.80	53.892	177.162	845.931
2.00	59.880	168.180	803.043
2.20	65.868	158.611	757.352
2.40	71.856	149.030	711.605
2.60	77.844	139.449	665.857
2.80	83.832	130.251	621.939
3.00	89.820	121.269	579.051
3.20	95.808	112.868	538.936
3.40	101.796	104.772	500.279
3.60	107.784	97.347	464.825
3.80	113.772	90.488	432.072
4.00	119.760	83.961	400.907
4.20	125.749	78.296	373.858
4.40	131.737	72.893	348.059
4.60	137.725	68.043	324.900
4.80	143.713	63.601	303.690
5.00	149.701	59.410	283.675
5.20	155.689	55.844	266.649
5.40	161.677	52.428	250.340
5.60	167.665	49.314	235.472
5.80	173.653	46.456	221.825
6.00	179.641	43.762	208.958
6.20	185.629	41.405	197.704
6.40	191.617	39.150	186.939
6.60	197.605	37.114	177.218
6.80	203.593	35.222	168.183
7.00	209.581	33.426	159.605
7.20	215.569	31.796	151.825
7.40	221.557	30.242	144.402
7.60	227.545	28.865	137.826
7.80	233.533	27.558	131.587

Contd...

8.00	239.521	26.301	125.583
8.20	245.509	25.153	120.105
8.350	250.000	24.300	116.030

Table XIV (17) Axial Compression Load Capacity of Channel-Section

ISLT 50 @ 4.0 kg/m

Section	*Area mm²*	$r_{min} = r_{yy}$ *mm*	*Buckling Class y-y*
ISLT 50 @ 4.0 kg/m	*510.548*	*11.20*	*c*
Eff Length m	*Sl Ratio*	*D.C.S.* f_{cd} *N/mm²*	*F.A.L. kN*
1.00	89.286	122.071	62.323
1.10	98.214	109.500	55.905
1.20	107.143	98.143	50.107
1.30	116.071	87.982	44.919
1.40	125.000	79.000	40.333
1.50	133.929	71.118	36.309
1.60	142.857	64.200	32.777
1.70	151.786	58.146	29.687
1.80	160.714	52.929	27.023
1.90	169.643	48.286	24.652
2.00	178.571	44.243	22.588
2.10	187.500	40.675	20.767
2.20	196.429	37.514	19.153
2.30	205.357	34.693	17.712
2.40	214.286	32.143	16.410
2.50	223.214	29.861	15.245
2.60	232.143	27.850	14.219
2.70	241.071	25.996	13.272
2.800	250.000	24.300	12.406

Table XIV (18) Axial Compression Load Capacity of Channel-Section

ISLT 75 @ 7.1 kg/m

Section	*Area mm²*	$r_{min} = r_{yy}$ *mm*	*Buckling Class y-y*
ISLT 75 @ 7.1 kg/m	*904.160*	*17.50*	*c*
Eff Length m	*Sl Ratio*	*D.C.S.* f_{cd} *N/mm²*	*F.A.L. kN*
1.00	57.143	172.286	155.774
1.10	62.857	163.429	147.766
1.20	68.571	154.286	139.499
1.30	74.286	145.143	131.232
1.40	80.000	136.000	122.966
1.50	85.714	127.429	115.216
1.60	91.429	119.000	107.595
1.70	97.143	111.000	100.362
1.80	102.857	103.457	93.542
1.90	108.571	96.371	87.135
2.00	114.286	89.929	81.310

Contd...

2.10	120.000	83.700	75.678
2.20	125.714	78.329	70.822
2.30	131.429	73.143	66.133
2.40	137.143	68.514	61.948
2.50	142.857	64.200	58.047
2.60	148.571	60.200	54.430
2.70	154.286	56.671	51.240
2.80	160.000	53.300	48.192
2.90	165.714	50.329	45.505
3.00	171.429	47.457	42.909
3.10	177.143	44.886	40.584
3.20	182.857	42.486	38.414
3.30	188.571	40.257	36.399
3.40	194.286	38.243	34.578
3.50	200.000	36.300	32.821
3.60	205.714	34.586	31.271
3.70	211.429	32.914	29.760
3.80	217.143	31.371	28.365
3.90	222.857	29.943	27.073
4.00	228.571	28.629	25.885
4.10	234.286	27.400	24.774
4.20	240.000	26.200	23.689
4.30	245.714	25.114	22.707
4.375	250.000	24.300	21.971

able XIV (19) Axial Compression Load Capacity of Channel-Section

ISLT 100 @ 12.7 kg/m

Section	*Area mm²*	*$r_{min} = r_{yy}$ mm*	*Buckling Class y-y*
ISLT 100 @ 12.7 kg/m	*1616.337*	*21.50*	*c*
Eff Length m	*Sl Ratio*	*D.C.S. f_{cd} N/mm²*	*F.A.L. kN*
1.00	46.512	188.233	304.247
1.20	55.814	174.279	281.694
1.40	65.116	159.814	258.313
1.60	74.419	144.930	234.256
1.80	83.721	130.419	210.800
2.00	93.023	116.767	188.736
2.20	102.326	104.116	168.287
2.40	111.628	92.826	150.037
2.60	120.930	82.826	133.874
2.80	130.233	74.112	119.789
3.00	139.535	66.577	107.610
3.20	148.837	60.014	97.003
3.40	158.140	54.398	87.925
3.60	167.442	49.430	79.896
3.80	176.744	45.065	72.840

Contd...

4.00	186.047	41.242	66.661
4.20	195.349	37.881	61.229
4.40	204.651	34.905	56.418
4.60	213.953	32.233	52.099
4.80	223.256	29.851	48.250
5.00	232.558	27.763	44.874
5.20	241.860	25.847	41.777
5.375	250.000	24.300	39.277

Table XIV (20) Axial Compression Load Capacity of Channel-Section

ISJT 75 @ 3.5 kg/m

Section	*Area mm²*	$r_{min} = r_{yy}$ *mm*	*Buckling Class y-y*
ISJT 75 @ 3.5 kg/m	*450.384*	*10.10*	*c*
Eff Length m	*Sl Ratio*	*D.C.S.* f_{cd} *N/mm²*	*F.A.L. kN*
1.00	99.010	108.386	48.815
1.10	108.911	95.950	43.215
1.20	118.812	84.995	38.280
1.30	128.713	75.510	34.008
1.40	138.614	67.323	30.321
1.50	148.515	60.240	27.131
1.60	158.416	54.235	24.426
1.70	168.317	48.975	22.058
1.80	178.218	44.402	19.998
1.90	188.119	40.434	18.211
2.00	198.020	36.973	16.652
2.10	207.921	33.924	15.279
2.20	217.822	31.188	14.047
2.30	227.723	28.824	12.982
2.40	237.624	26.699	12.025
2.50	247.525	24.770	11.156
2.525	250.000	24.300	10.944

Table XIV (21) Axial Compression Load Capacity of Channel-Section

ISJT 87.5 @ 4.0 kg/m

Section	*Area mm²*	$r_{min} = r_{yy}$ *mm*	*Buckling Class y-y*
ISJT 87.5 @ 4.0 kg/m	*513.824*	*9.70*	*c*
Eff Length m	*Sl Ratio*	*D.C.S.* f_{cd} *N/mm²*	*F.A.L. kN*
1.00	103.093	103.165	53.009
1.10	113.402	90.892	46.702
1.20	123.711	80.211	41.215
1.30	134.021	71.043	36.504
1.40	144.330	63.169	32.458
1.50	154.639	56.463	29.012
1.60	164.948	50.727	26.065
1.70	175.258	45.734	23.499

Contd...

1.80	185.567	41.429	21.287
1.90	195.876	37.702	19.372
2.00	206.186	34.444	17.698
2.10	216.495	31.546	16.209
2.20	226.804	29.035	14.919
2.30	237.113	26.806	13.774
2.40	247.423	24.790	12.738
2.425	250.000	24.300	12.486

able XIV (22) Axial Compression Load Capacity of Channel-Section

ISJT 100 @ 5.0 kg/m

Section	Area mm^2	$r_{min} = r_{yy}$ mm	Buckling Class y-y
ISJT 100 @ 5.0 kg/m	632.184	*11.70*	c
Eff Length m	Sl Ratio	D.C.S. f_{cd} N/mm^2	F.A.L. kN
1.00	85.470	127.795	80.790
1.10	94.017	115.376	72.939
1.20	102.564	103.821	65.634
1.30	111.111	93.389	59.039
1.40	119.658	84.073	53.149
1.50	128.205	75.987	48.038
1.60	136.752	68.831	43.514
1.70	145.299	62.491	39.506
1.80	153.846	56.931	35.991
1.90	162.393	52.056	32.909
2.00	170.940	47.677	30.141
2.10	179.487	43.831	27.709
2.20	188.034	40.467	25.582
2.30	196.581	37.462	23.683
2.40	205.128	34.762	21.976
2.50	213.675	32.308	20.424
2.60	222.222	30.089	19.022
2.70	230.769	28.138	17.789
2.80	239.316	26.344	16.654
2.90	247.863	24.706	15.619
2.925	250.000	24.300	15.362

able XIV (23) Axial Compression Load Capacity of Channel-Section

ISJT 112.5 @ 6.4 kg/m

Section	Area mm^2	$r_{min} = r_{yy}$ mm	Buckling Class y-y
ISJT 112.5 @ 6.4 kg/m	813.898	*15.80*	c
Eff Length m	Sl Ratio	D.C.S. f_{cd} N/mm^2	F.A.L. kN
1.00	63.291	162.734	132.449
1.10	69.620	152.608	124.207
1.20	75.949	142.481	115.965
1.30	82.278	132.582	107.908

Contd...

1.40	88.608	123.089	100.182
1.50	94.937	114.089	92.856
1.60	101.266	105.430	85.810
1.70	107.595	97.582	79.422
1.80	113.924	90.323	73.514
1.90	120.253	83.462	67.930
2.00	126.582	77.513	63.087
2.10	132.911	71.942	58.553
2.20	139.241	66.815	54.381
2.30	145.570	62.301	50.707
2.40	151.899	58.080	47.271
2.50	158.228	54.346	44.232
2.60	164.557	50.930	41.452
2.70	170.886	47.701	38.824
2.80	177.215	44.853	36.506
2.90	183.544	42.218	34.361
3.00	189.873	39.749	32.352
3.10	196.203	37.591	30.595
3.20	202.532	35.541	28.926
3.30	208.861	33.642	27.381
3.40	215.190	31.899	25.962
3.50	221.519	30.251	24.621
3.60	227.848	28.795	23.436
3.70	234.177	27.423	22.319
3.80	240.506	26.104	21.246
3.90	246.835	24.901	20.267
3.95	250.000	24.300	19.778

84	β 0.1024	β 0.1296	β 0.16	β 0.1936	β 0.2304	β 0.27	β 0.3136	β 0.36	β 0.4096	β 0.4624	β 0.5184	β 0.5776	β 0.64
n)	M_{dV} (kNm)	M_{dV} (kNm)	M_{dV} (kNm)	M_{dV} (kNm)	M_{dV} (kNm)	M_{dV} (kNm)	M_{dV} (kNm)	M_{dV} (kNm)	M_{dV} (kNm)	M_{dV} (kNm)	M_{dV} (kNm)	M_{dV} (kNm)	M_{dV} (kNm)
	10.874	10.770	10.653	10.524	10.383	10.230	10.064	9.886	9.696	9.493	9.279	9.052	8.812
	14.026	13.875	13.705	13.518	13.313	13.091	12.850	12.592	12.315	12.021	11.710	11.380	11.033
	19.867	19.657	19.422	19.163	18.878	18.569	18.235	17.877	17.494	17.086	16.653	16.195	15.713
	29.401	29.111	28.788	28.430	28.039	27.613	27.153	26.659	26.131	25.570	24.974	24.343	23.679
	4.959	4.927	4.891	4.851	4.808	4.760	4.709	4.654	4.596	4.533	4.467	4.397	4.323
	8.605	8.543	8.474	8.398	8.314	8.223	8.125	8.020	7.907	7.787	7.660	7.525	7.383
	16.403	16.296	16.178	16.046	15.903	15.746	15.578	15.396	15.203	14.996	14.778	14.546	14.303
	23.123	22.957	22.770	22.564	22.338	22.093	21.827	21.543	21.238	20.914	20.571	20.207	19.825
	31.502	31.261	30.991	30.693	30.366	30.011	29.628	29.216	28.776	28.308	27.811	27.285	26.731
	42.595	42.261	41.888	41.476	41.024	40.533	40.003	39.433	38.825	38.177	37.489	36.763	35.997
	56.178	55.724	55.217	54.657	54.043	53.375	52.655	51.881	51.053	50.172	49.238	48.250	47.209
	74.753	74.163	73.505	72.777	71.980	71.113	70.178	69.172	68.098	66.954	65.741	64.459	63.107
	97.999	97.251	96.415	95.491	94.479	93.379	92.191	90.915	89.551	88.099	86.559	84.931	83.215
	122.469	121.537	120.496	119.345	118.084	116.713	115.233	113.644	111.944	110.135	108.217	106.188	104.050
	152.000	150.857	149.580	148.168	146.622	144.942	143.127	141.178	139.094	136.876	134.524	132.037	129.415
	188.157	186.756	185.190	183.459	181.564	179.504	177.279	174.889	172.334	169.615	166.730	163.681	160.467
	242.426	240.443	238.237	235.793	233.117	230.208	227.066	223.691	220.084	216.244	212.171	207.866	203.328
	308.342	305.650	302.642	299.317	295.676	291.718	287.444	282.852	277.944	272.720	267.179	261.321	255.147
	389.495	385.940	381.967	377.577	372.767	367.540	361.895	355.831	349.349	342.449	335.131	327.395	319.240
	488.978	484.349	479.177	473.459	467.198	460.391	453.041	445.145	436.706	427.721	418.193	408.119	397.502
	614.037	608.195	601.666	594.449	586.546	577.955	568.677	558.711	548.058	536.718	524.691	511.977	498.575
	13.097	13.035	12.966	12.889	12.806	12.715	12.617	12.511	12.399	12.279	12.151	12.017	11.875
	18.201	18.095	17.976	17.845	17.701	17.545	17.376	17.195	17.001	16.795	16.576	16.345	16.101
	24.480	24.313	24.126	23.920	23.694	23.449	23.184	22.899	22.595	22.271	21.927	21.564	21.181
	36.765	36.504	36.213	35.892	35.540	35.157	34.743	34.299	33.825	33.319	32.783	32.217	31.620
	56.368	56.016	55.622	55.186	54.710	54.192	53.632	53.031	52.388	51.704	50.979	50.212	49.404
	77.238	76.729	76.161	75.533	74.845	74.097	73.289	72.422	71.494	70.507	69.460	68.353	67.187
	103.335	102.668	101.923	101.100	100.198	99.218	98.160	97.023	95.807	94.514	93.142	91.691	90.162
	144.193	143.150	141.984	140.696	139.284	137.750	136.093	134.314	132.412	130.387	128.239	125.968	123.575
	196.403	194.870	193.156	191.262	189.187	186.932	184.496	181.880	179.084	176.107	172.950	169.613	166.095
	259.025	256.824	254.364	251.646	248.668	245.432	241.937	238.183	234.169	229.897	225.367	220.577	215.528
	341.959	339.017	335.729	332.095	328.115	323.789	319.117	314.098	308.734	303.023	296.967	290.564	283.815
	456.675	452.734	448.330	443.461	438.130	432.334	426.075	419.352	412.166	404.516	396.402	387.825	378.784
	596.642	591.406	585.554	579.086	572.002	564.302	555.986	547.054	537.506	527.342	516.562	505.166	493.154
	772.729	766.052	758.590	750.343	741.310	731.492	720.889	709.500	697.325	684.365	670.620	656.089	640.772
	28.124	27.936	27.727	27.495	27.241	26.964	26.666	26.346	26.004	25.639	25.252	24.844	24.413
	43.101	42.826	42.519	42.180	41.809	41.405	40.969	40.501	40.000	39.467	38.902	38.305	37.675
	65.395	65.018	64.597	64.131	63.621	63.066	62.467	61.824	61.136	60.404	59.628	58.807	57.942
	86.736	86.235	85.675	85.057	84.379	83.643	82.848	81.993	81.080	80.108	79.077	77.986	76.839
	220.545	219.030	217.337	215.466	213.417	211.190	208.785	206.201	203.439	200.499	197.381	194.085	190.610
	292.034	289.908	287.531	284.904	282.027	278.900	275.522	271.895	268.017	263.889	259.511	254.882	250.004
	389.292	386.412	383.195	379.638	375.743	371.508	366.936	362.024	356.774	351.185	345.257	338.991	332.385
	520.452	516.627	512.352	507.627	502.452	496.827	490.752	484.227	477.252	469.827	461.952	453.627	444.852
	678.396	673.487	668.001	661.937	655.296	648.077	640.281	631.907	622.956	613.427	603.321	592.637	581.376
	882.596	876.365	869.401	861.703	853.273	844.109	834.212	823.583	812.220	800.124	787.295	773.732	759.437
	963.862	957.297	949.959	941.849	932.967	923.313	912.886	901.687	889.715	876.971	863.455	849.166	834.105

Table XVI Moments of Resistance of Semi-Compact I-Sections

Semi Compact Sections	*MR (kNm)*
ISWB250 @40.9kg/m	108.0455
ISHB200 @ 37.3kg/m	82.0000
ISHB200 @40.0kg/m	84.5909
ISHB225 @43.1kg/m	106.6591
ISHB225 @46.8kg/m	110.6818
ISHB250 @51.0kg/m	140.6591
ISHB250 @54.7kg/m	145.1591
ISHB300 @58.8kg/m	190.0682
ISHB300 @63.0kg/m	196.2045
ISHB350 @67.4kg/m	248.8182
ISHB350 @72.4kg/m	257.1818

Annexure

Annexure A-1 Shear Strength of Grade 4.6 Bolts

Diameter (mm)	Shank Area (mm²)	Net Area (mm²)	Single Shear (kN)	Double Shear (kN)
12	113.10	88.22	16.298	37.193
14	153.94	120.07	22.183	50.624
16	201.06	156.83	28.974	66.121
18	254.47	198.49	36.671	83.684
20	314.16	245.04	45.272	103.314
22	380.13	296.50	54.780	125.010
24	452.39	352.86	65.192	148.772
27	572.56	446.59	82.509	188.290
30	706.86	551.35	101.863	232.457
33	855.30	667.13	123.254	281.272
36	1017.88	793.94	146.683	334.737

Annexure A-2 Tensile Strength of Grade 4.6 Bolts

Diameter (mm)	In Yeilding (kN)	At Ult.Stress (kN)
12	24.676	25.406
14	33.586	34.581
16	43.868	45.167
18	55.521	57.164
20	68.544	70.573
22	82.938	85.393
24	98.703	101.625
27	124.921	128.619
30	154.224	158.789
33	186.611	192.134
36	222.082	228.656

Annexure A-3

Diameter d (mm)	Bolt Hole Dia. do (mm)	Min. End Dist. e (mm)	Min. Pitch, p (mm)	Suggested min. e	Suggested min. p	Suggested kb	Suggested Bearing Strength (kN)	t = 6	t = 8	t = 10	t = 12	t=14	t = 16	t = 18	t = 20	t = 22	t = 24
12	13	19.5	30	20	30	0.5128	4.9231*t	29.538	39.385	49.231	59.077	68.923	78.769	88.615	98.462	108.308	118.154
14	15	22.5	35	25	35	0.5278	5.9111*t	35.467	47.289	59.111	70.933	82.756	94.578	106.400	118.222	130.044	141.867
16	18	27	40	30	40	0.4907	6.2815*t	37.689	50.252	62.815	75.378	87.941	100.504	113.067	125.630	138.193	150.756
18	20	30	45	30	45	0.5000	7.2000*t	43.200	57.600	72.000	86.400	100.800	115.200	129.600	144.000	158.400	172.800
20	22	33	50	35	50	0.5076	8.1212*t	48.727	64.970	81.212	97.455	113.697	129.939	146.182	162.424	178.667	194.909
22	24	36	55	40	55	0.5139	9.0444*t	54.267	72.356	90.444	108.533	126.622	144.711	162.800	180.889	198.978	217.067
24	26	39	60	40	60	0.5128	9.8462*t	59.077	78.769	98.462	118.154	137.846	157.538	177.231	196.923	216.615	236.308
27	30	45	67.5	45	70	0.5000	10.8000*t	64.800	86.400	108.000	129.600	151.200	172.800	194.400	216.000	237.600	259.200
30	33	49.5	75	50	75	0.5051	12.1212*t	72.727	96.970	121.212	145.455	169.697	193.939	218.182	242.424	266.667	290.909
33	36	54	82.5	55	85	0.5093	13.4444*t	80.667	107.556	134.444	161.333	188.222	215.111	242.000	268.889	295.778	322.667
36	39	58.5	90	60	90	0.5128	14.7692*t	88.615	118.154	147.692	177.231	206.769	236.308	265.846	295.385	324.923	354.462
12	13	19.5	30	20	35	0.5128	4.9231*t	29.538	39.385	49.231	59.077	68.923	78.769	88.615	98.462	108.308	118.154
14	15	22.5	35	25	40	0.5556	6.2222*t	37.333	49.778	62.222	74.667	87.111	99.556	112.000	124.444	136.889	149.333
16	18	27	40	30	45	0.5556	7.1111*t	42.667	56.889	71.111	85.333	99.556	113.778	128.000	142.222	156.444	170.667
18	20	30	45	30	50	0.5000	7.2000*t	43.200	57.600	72.000	86.400	100.800	115.200	129.600	144.000	158.400	172.800
20	22	33	50	35	55	0.5303	8.4848*t	50.909	67.879	84.848	101.818	118.788	135.758	152.727	169.697	186.667	203.636
22	24	36	55	40	60	0.5556	9.7778*t	58.667	78.222	97.778	117.333	136.889	156.444	176.000	195.556	215.111	234.667
24	26	39	60	40	65	0.5128	9.8462*t	59.077	78.769	98.462	118.154	137.846	157.538	177.231	196.923	216.615	236.308
27	30	45	67.5	45	70	0.5000	10.8000*t	64.800	86.400	108.000	129.600	151.200	172.800	194.400	216.000	237.600	259.200
30	33	49.5	75	50	80	0.5051	12.1212*t	72.727	96.970	121.212	145.455	169.697	193.939	218.182	242.424	266.667	290.909
33	36	54	82.5	55	85	0.5093	13.4444*t	80.667	107.556	134.444	161.333	188.222	215.111	242.000	268.889	295.778	322.667
36	39	58.5	90	60	95	0.5128	14.7692*t	88.615	118.154	147.692	177.231	206.769	236.308	265.846	295.385	324.923	354.462

Bearing Strength of Grade 4.6 Bolts for Typical End Distances and Pitch Distances

Annexure A-4

	Shop Weld (N/mm²)	189.3709	
	Field Weld (N/mm²)	157.8091	
Size, s (mm)	Throat th., t (mm)	Unit Strength (kN/mm)	
		Shop	Field
3.00	2.10	0.3977	0.3314
3.50	2.45	0.4640	0.3866
4.00	2.80	0.5302	0.4419
4.50	3.15	0.5965	0.4971
5.00	3.50	0.6628	0.5523
5.50	3.85	0.7291	0.6076
6.00	4.20	0.7954	0.6628
6.50	4.55	0.8616	0.7180
7.00	4.90	0.9279	0.7733
7.50	5.25	0.9942	0.8285
8.00	5.60	1.0605	0.8837
8.50	5.95	1.1268	0.9390
9.00	6.30	1.1930	0.9942
9.50	6.65	1.2593	1.0494
10.00	7.00	1.3256	1.1047
10.50	7.35	1.3919	1.1599
11.00	7.70	1.4582	1.2151
11.50	8.05	1.5244	1.2704
12.00	8.40	1.5907	1.3256

Fillet Weld Strength for Shop Welds and Field welds